Muhammed Abed Mazeel

Iraq Oil and Gas Papers 2011

disserta
Verlag

Mazeel, Muhammed Abed: Iraq Oil and Gas Papers 2011, Hamburg, disserta Verlag, 2012

ISBN: 978-3-95425-002-8
Druck: disserta Verlag, Hamburg, 2012

Bibliografische Information der Deutschen Nationalbibliothek
Die Deutsche Nationalbibliothek verzeichnet diese Publikation in der Deutschen Nationalbibliografie; detaillierte bibliografische Daten sind im Internet über http://dnb.d-nb.de abrufbar.

Die digitale Ausgabe (eBook-Ausgabe) dieses Titels trägt die ISBN 978-3-95425-003-5 und kann über den Handel oder den Verlag bezogen werden.

Published by disserta Verlag, Hamburg, Germany, 2012

Table of Contents

Figures and Tables

See the articles

The revenue from this book will be donated to the sick cancer children and help organizations.

Dr Muhammed Mazeel
29.01.2012

1. About the Book

This book contains my articles from 2011. I have tried to examine some of the technical problems in the current oil and gas fields/prospects in Iraq. In the geology/geophysics and reservoir engineering section, I used the available information and in some cases, I estimate to fill the gap in the existing data

Iraq Oil Ministry has offered twelve areas covering 81,700 square kilometres with the blocks having a mix of oil and gas productivity. However, many of the blocks contain oil and gas shows and some of them very close to previously discovered oil and gas fields. The Western and South western desert is estimated to have high potential for hydrocarbon, where many thousand of kilometres of 2D seismic data were acquired and the own state Oil Exploration Company (OEC) has discovered several fields.

The mentioned regions are thought to be highly prospective and contain numerous undrilled prospects. Only four exploration wells have been drilled on the blocks and the seismic coverage is old and of poor quality, this include Western and South western to the South of the country. The potential of discovering high potential reservoir is very considerable which may change the Iraqi total reserves to 300 or more billion barrel of oil and 500 trillion bscf of gas.

The objective of this book (published articles) is to identify and evaluate the petroleum systems of prospects and leads, calculate the reserves and evaluate the hydrocarbon potential and economics of the 12 blocks, Akkas and Ahdab fields using existing literature and available data published by the Ministry of Oil.

2. Hydrocarbon potential, reservoir development estimated for western Iraq's Akkas gas field

Akkas is in northwestern Iraq near the border with Syria. It is 52 km from the T1 pump station and 285 km from the K3 pump station. The distance between this field and Baji town, where the entire Iraqi gas grid passes, is around 300 km. The nearest industrial complex to the field is the Al-Qaim Industrial Complex that consists of a fertilizer plant and cement plant and is less than 40 km away. The first exploration well in Akkas region took place in August 1992. The drilling program target was to reach a depth of 5,000 m, but this did not succeed due to technical difficulties. The drilling reached a depth of 4,238 m, and the well test confirmed the presence of natural gas with a flow rate of 6-8 MMscfd. Akkas is one of the largest gas fields in Iraq. It contains 5.68 tscf of initial gas in place, of which 4.55 tscf is estimated to be recoverable. There is also the potential of condensate and other prospects in deeper formations. In 2001, the development of Akkas-Saladin gas field was referred to the Syrian Petroleum Co. The commitment was to drill five horizontal wells and redrill the SA-1 (Akkas-1) discovery well horizontally. An average well test for another five wells is demonstrating more than 9 MMscfd on the largest choke.

Geology and Petrophysics

The formations with hydrocarbon potential in Akkas field are as follows:

- The Ora Kaista Pirispiki formation at 1,365-1,424 m is a sequence of sandstone, a compact shale layer of low and medium porosity, followed by dolomite and limestone.
- The Akkas formation at 1,993-2,002 m is sandstone with compact shale layers of low porosity.
- The Khabour formation, at 2,332-60 m in Akkas-1 (SA-1), 2,365-75.5 m in Akkas-2 (SA-2), and 2,341-55 m in Akkas-3 (SA-3), is sandstone and compact shale layers of low porosity. The specific gravity of the tested gas was 0.726-0.6953 and of the condensate was 0.7792.

In Akkas-1 (SA-1) are indications of light oil (density 0.8326 gm/cu cm). An Akkas-4 (SA-4) test showed gas, and the author believes condensate may also be present in this well. The gas fields of the Widyan basin-Interior platform in western Iraq are mainly Lower Paleozoic petroleum systems such as gas in the Lower Silurian Akkas and Ordovician Khabour formations. Their main source could be the Khabour formation and the oil shale of the Akkas formation. Structural framework of folding and faulting as well as stratigraphic facies changes could form the main traps in these regions. Samples of cores and cuttings were collected from the Khabour, Akkas, and Upper Devonian Kaista formations in wells Akkas 1-6, Khleisya-1, KH 5-6, and KH 5-1. Their diagnostic organic matters are abundant and a few spores and chitinozoa as well as scolecodonts, graptolite siculae, cuticles, and amorphous organic matter.

Hydrocarbon generation potential is assessed by plotting organic matter up to 16% TOC, especially the hot shale of the Akkas formation, which is very low asphaltene and sulfur. The saturated and aromatic hydrocarbons of more than 96% and high peaks of C_2-C_{20} gas chromatography could indicate predominant gas generation with some light oils. The associated gases are mainly methane and ethane of CH_4, C_4H_6, and C_3H_8. Accordingly, source potential for wet gas and condensate could be assessed at a depth of 2,750-3,000 m and dry gas at a depth of 3,570-3,650 m in Akkas-1 only from the Khabour formation. Little oil might be generated from the Akkas formation in Akkas-1 and KH 5-6 (Fig. 1).

AKKAS-1 (SA-1) STRATIGRAPHIC COLUMN

FIG. 1

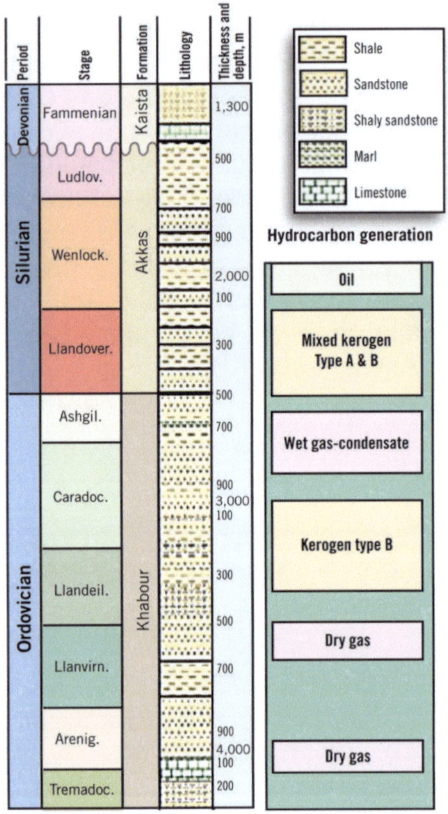

Source: Senso Bujak et al., 1977

These potential source rocks are extended toward neighbor countries. Accumulation sites of these generated gases and a little oil could be within the sandstone porosities of 10-17% and permeability of 500 md sealed by the nonpermeable shale along closures of the structured anticline fold and fault of this field as well as along the unconformity of the boundary of the Akkas formation with the Kaista formation. Accordingly, Lower Paleozoic Total Petroleum System of generation, migration, and accumulation could be assessed for a basin includes West Iraq and extensions in the neighboring regions.

Formation Evaluation

In Akkas-1, the borehole generated condensates as well as wet and dry gas of mainly 85% methane and ethane. Little oil could have been generated from the upper part of the Lower Silurian rocks. Increasing thermal alteration (>170° C., TAI = 3.8) applied to rocks containing more than 0.5% TOC would be a reason for generating gas from Ordovician rocks. Trapped gas and oil could be accumulated along anticline and fault structural traps within Ordovician and Silurian sandstone's interlayers in West Iraq. Table 1 shows summarized characteristics of the Khabour formation obtained from the tests made in the Akkas-1 (SA-1) well:

PUBLISHED DATA FROM AKKAS-1 (SA-1) WELL Table 1

Gas initially in place	5.680 tcf
Condensate-gas ratio	Condensate-gas ratio
Cond.-gas sp gr	0.779
Cond.-gas gravity	50° API
Cond. water gas ratio	0.20%
Cond. hydrogen sulfide	0.06
Gas sp gr	0.726
Carbon dioxide	0.04
Nitrogen	0
Water salinity	120,000
Gas density at 1 atm	0.89
Reservoir temperature	210° F.
Reservoir pressure	3,720 psi
Flow rate	17.3 MMscfd
Flow pressure	3,045 psi
Choke size	0.5 in.
Wellhead pressure	2,015 psi
Pressure at separator	1,100 psi
Choke temperature	135° F.
Pipe temperature	130° F.
Survey depth	2,100/3,000/1,000 m
Pipe length to separator	1,000 m
Tubing depth	2,150 is 3.5 in.

Wet gas and condensate generation

The Akkas-1 borehole has total organic carbon of 0.71-1.42 wt % at 2,750-3,000 m. The biodegradation and thermal alteration of the organic matter resulted in abundant amorphous organic matter (70-75%).The Khabour B rock unit started the generation of wet gas and condensates hydrocarbons during the Mississippian (Mesopotamian) time and continued till now with the expulsion quantity of 65% of its proven reserve.

Dry Gas Generation

The Akkas-1 borehole has TOC of 0.5-1.0% at 3,570-3,650 m. The biodegradation and thermal alteration of the organic matter led to abundant amorphous organic matter. The analyses for this shale unit (Khabour D) is giving results of starting dry gas generation during Silurian time and ended generation during late Triassic.

Oil Generation

Potential source rocks for oil have been encountered from the Akkas formation depths of 2,280-2,330 m in Akkas-1. The analyses for this lithostratigraphic unit of Akkas hot shale have resulted in no hydrocarbon generation although it has very high organic content.

Subsurface Condition

Low sulfur content of the generated hydrocarbons in Akkas field could be explained by the presence of iron in these clastic marine rocks that combines with sulfur and precipitates as iron pyrites (which is found in abundance with the organic kerogen). The free sulfur content is reduced in the organic matter and in the released hydrocarbons. In addition, an increase in the geothermal gradient due to the presence of radioactive shales has caused the breakdown of long chain hydrocarbon compounds into simpler compounds forming very light oil, condensate, and gas with low sulfur contents. This was proven by the test results from Akkas-1, which has 42° gravity light oil in sandstone units at the base of Silurian and free sweet gas in the upper part of the Khabour formation. The results of bitumen analysis from the shale source rock unit, at the base of Silurian, indicated that saturated hydrocarbons and aromatics make up 96 wt % and asphalt compounds 3.89 wt % of the extract. The saturated compounds are of low molecular weight (C_2-C_{20}) by gas chromatography and hence could indicate predominant gas generation with some light oils.

Reservoir and Seal

This principal reservoir (Khabour formation) is the oldest known sedimentary rock unit in Iraq. It consists entirely of siliciclastic sequences, comprising thin-bedded, fine-grained sandstone, quartzite graywackes, and silty micaceous shale with some intercalation of dolomite and limestone. The base of the formation, which may be more than 1,000 m thick, has not been reached in wells. It was deposited in shallow marine inner to outer neritic environments that prevailed over the entire eastern part of the Arabian Plate. In Akkas field, the Khabour formation is found at depths below 2,310 m and exhibits good reservoir properties. It comprises sandstone with silty shales and has a gross thickness of approximately 38 m and an average thickness of 25 m. Core analysis of the reservoir intervals generally indicated an average porosity of 10% and permeability of 500 md. Its depth in Akkas-1 is at 2,040 m below sea level. Formation temperature is 146° C. at 3,300 m. The high temperature is caused by the Silurian shales' radioactivity and decreases downward in the Khabour formation to normal gradients. The gross hydrocarbon reservoir column is about 80 m. A secondary reservoir is provided by the Akkas formation found above the Khabour formation at a drilled depth of 1,465-2,326 m. It is comprised of a sandstone unit inter-bedded within the basal hot shale unit, with a gross thickness of 10 m and net thickness of 1.5 m. Core analysis of the reservoir intervals indicated porosity of 17% and a permeability of 500 md. The Akkas formation was found to contain 42° gravity light oil only from the thin sandstone horizon sandwiched between the hot shales at the base; it is therefore considered a minor reservoir. The overlying Silurian shale of the Akkas formation is an effective local regional seal.

Reservoir Engineering

The author's MBAL material balance model indicates that cumulative recoverable gas at Akkas would be 3.553 tscf. A possible production profile plateau is 500 MMscfd from 50 wells and for contract reason can be 55 wells. Published data show many uncertainties in the reservoir like bottomhole pressure, wellhead pressure, density, and downhole equipment. From published information it is impossible to calculate the vertical lift performance (VLP) and the inflow performance relationship (IPR). The author estimated the wellhead pressure according to the given reservoir pressure of 3,720 psi. The author used available data and estimated the well data, which are not available or available but inconsistent with reservoir data. In order to design and allow for the building of reliable and consistent well models with the ability to address each aspect of wellbore modeling, PVT (fluid characterization), vertical lift performance (VLP) correlations (for calculation of flow line and tubing pressure loss) and IPR (reservoir inflow), the author used Prosper.

ABSOLUTE OPEN FLOW, IPR PLOT MULTIRATE C AND N

FIG. 2

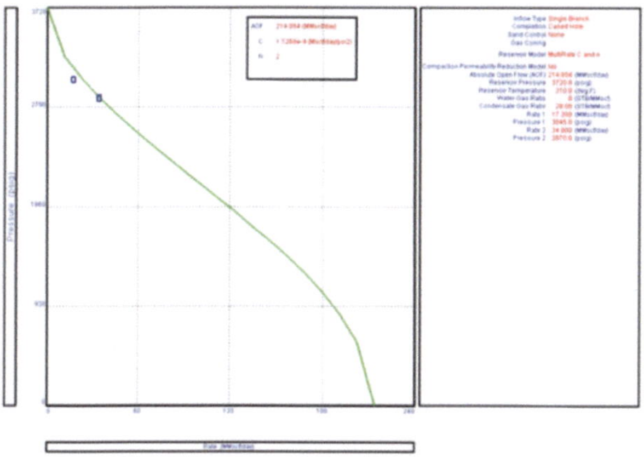

According to the available data from Akkas-1, the Prosper model shows the absolute open flow of 214 MMscfd (Fig. 2). It shows VLP and inflow performance relationship (IPR) with production rates of 17.4 MMscfd (Fig. 3).

INFLOW (IPR) VS. OUTFLOW (VLP)

FIG. 3

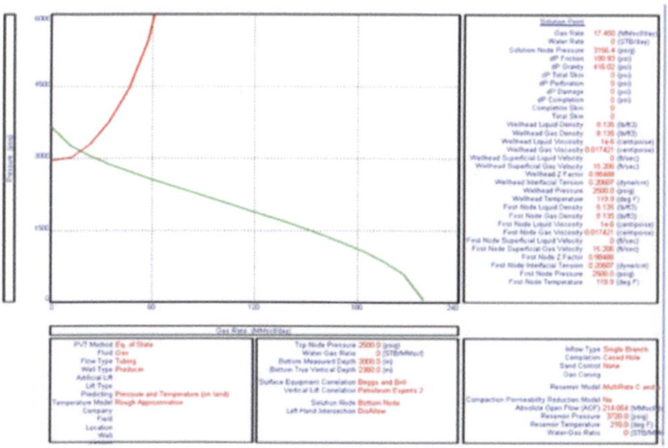

Efficient reservoir development requires a good understanding of reservoir and production systems. The author used MBAL to define reservoir drive mechanisms and hydrocarbon volumes and finally the reliable reservoir models quickly.

PRODUCTION PROFILE OF 50 AKKAS PRODUCTION WELLS

FIG. 4

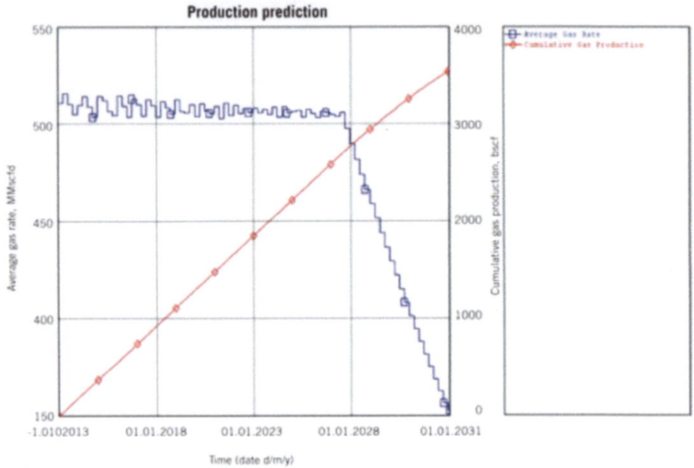

Fig. 4 shows the rough MBAL model for the tank Akkas-1 (SA-1) and extrapolated rates of the 50 wells including the five existing wells. According to the well test rates from the exploration wells, I take a conservative average production of 10 MMscfd/well.

Field Development Plan

Iraq proposed a strategy plan and wished to use the methane to satisfy domestic power/water demand, C_1, C_2, C_3 based petrochemicals for export. Ethane and propane are used for olefins production. Butane and condensates are for domestic use, export, and petrochemical production. The proposed development plan can be divided in two phases: The first phase is 250 MMscfd gas rate. Power generation and methanol use natural gas. Propane, butane, and condensate are for export. The second phase is an additional 250 MMscfd gas processed including ethane recovery for olefin production. Furthermore, power generation and ammonia use the incremented gas. Ethane is to be used for ethylene production, with the option to use some propane and butane. Polyethylene production is the main ethylene derivative.

Gas and Condensate Specs

The samples are taken from the separator (Tables 2 and 3).

SPECIFICATIONS OF FREE GAS FROM SA-1 AND SA-2 WELLS — Table 2

Component	SA-1	SA-2
	(2,332-60 m)	—
N_2	1.54	0
C_1	80.49	80.5299
C_2	8.92	8.9407
CO_2	2.2	2.0683
H_2S	NA	NA
C_3	3.54	3.3636
I-C_4	0.4	0.3788
n-C_4	1	0.9898
I-C_5	0.35	0.2924
n-C_5	0.41	0.3768
C_6	0.54	0.6784
C_7	0.39	0.7881
C_8	0.22	1.593
Sp gr	0.726	0.787
Density 15.5, 1AT-G/L	0.89	NA
Molecular weight	21.02	22.811
Heating value kcal/cu m:		
Gross	10,702	NA
Net	9,700	NA

SPECIFICATION OF AKKAS CONDENSATES — Table 3

Specification	SA-1	SA-2	SA-3
Sp gr	0.779	0.7538	0.7448
Water content, vol %	0.2	Nil	Nil
Sulfur content, wt %	0.06	0.0163	0.0131
Asphalt content	Nil	Trace	Nil
Pour point, °F.	–30	–45	–45
Kinematics viscosity (Cst) at:			
80° F.	1.52	0.98	1.016
100° F.	1.27	0.82	0.941
120° F.	1.08	0.75	0.876

Central Plant Facility

The proposed central plant facility is according to the possible development scenarios of the field and the use of the hydrocarbon and their derivates (Fig. 5).

POSSIBLE DESIGN OF AN AKKAS CENTRAL PRODUCTION FACILITY

FIG. 5

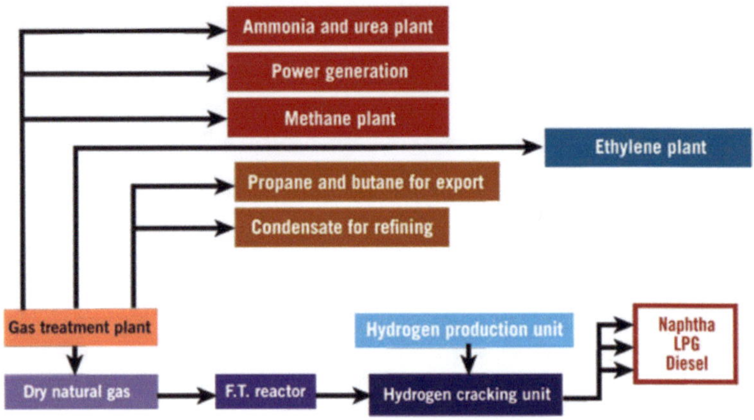

Pipelines and Route Connection

There are many possibilities to transport the gas and their products. The optimized way is to use the route of T1 to the Baniyas port in Syria (Fig. 6).

EXISTING ROUTE OF PIPELINES

FIG. 6

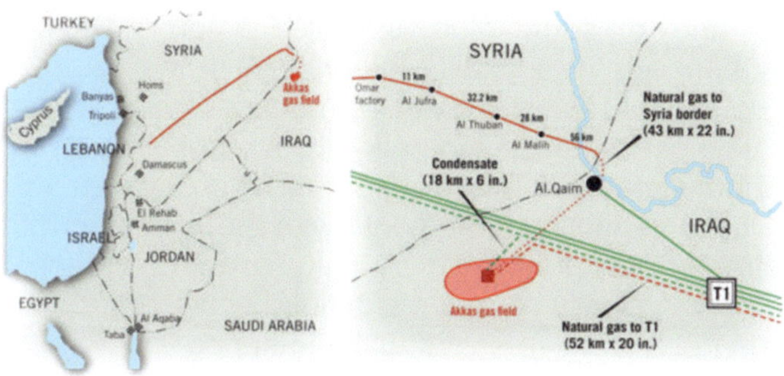

Possible Contract Conditions

Currently the main Iraq has only the service contracts, which still do not have the approval of political actors and public organs. The term service contracts encompass those various contracts in which the host government has a contract with a service company or an international oil company for the performance of services related to the exploitation of petroleum resources. The risk service contract appears similar to the production-sharing contract but differs in certain important matters. Its basic distinctive feature is that it reimburses the contractor in cash, not in crude oil or gas, although it may have provisions permitting the contractor to buy back an amount of crude oil at an international selling price equivalent to the amount to be paid to the contractor. A pure service contract is an agreement between a contractor and a host country that typically covers a defined technical service to be provided or completed during a specific period of time. The service company investment is typically limited to the value of equipment, tools, and personnel used to perform the service. In most cases, the service contractor's reimbursement is fixed by the terms of the contract with little exposure to either project performance or market factors. Payment for services is normally based on daily or hourly rates, a fixed rate, or some other specified amount. Payments may be made at specified intervals or at the completion of the service. Payments, in some cases, may be tied to the field performance, operating cost reductions, or other important metrics. These agreements are similar to the production-sharing agreements with the exception of the contractor payment. With a risk service contract, the contractor usually receives a defined share of production (in kind). As in the production-sharing contract, the contractor provides the capital and technical expertise required for exploration and development. If exploration efforts are successful, the contractor can recover those costs from the sale revenues and receive a share of profits through a contract-defined mechanism.

Cost Estimation

The approximate cost estimation for Akkas development phases are illustrated in Table 4. The calculations have been made for operating costs, based on the gas price for October 2010.

COSTS ESTIMATION OF PROPOSED AKKAS DEVELOPMENT Table 4

Seismic	G&G	G&A	Drilling	Facilities	Pipelines
			Million $		
14	5	10	40	70	18
14	5	10	80	80	
			120		
			120		
			40		

Workover rig rates will be around $20,000/day.The annual fixed operating costs for the central processing facility will be in range of $30-40 million/year. In addition, the fixed operating costs for the existing Iraqi pipelines will be $2 million/year. The variable operating costs are 20-30¢/bbl of oil or condensate, and 4.5-7.5¢/Mcf.

Author's Assumptions

This article sets out assumptions about Akkas development that the author has derived from poor published data. The inconsistency between the downhole pressure, reservoir pressure, and the wellhead pressure published numbers is solved by estimation. The possible production rate and plateau is based on a conservative production development scenario. Akkas field shows very high gas rate potential. Further steps should be taken:

- Analyze samples from Akkas wells to evaluate their biomarkers and thereby confirm the oil and gas sources in order to find other oil and gas pays in Akkas and oil and gas fields in the west of Iraq.
- Analysis of local oil seeps has been conducted to find their affinity and the structural relation to Akkas oil and gas fields. It is also possible that they lead to other oil fields that have been charged by oil migration from Mesopotamian basin to structural closures in West Iraq.
- It is important to drill deeper at least in the exploration/appraisal phases to find possible gas accumulations.

It is well known that the Iraq government would like to develop Akkas and is looking for new discoveries in the Western Desert of Iraq. This region has not yet been explored and may hold potential. The development plan for the use of Akkas gas is more likely to serve the national grid of electricity, future and existing petrochemicals, and other industrial plants. Part of the derivates will be exported. Most likely Iraq will follow the Saudi Aramco plan that uses the gas for development of national industry and demand. Those can multiply the profits from the petrochemicals and other industries of the petroleum resources. The Iraq government chooses the service contracts for different political and regulatory reasons (like delays or unapproved petroleum law). The long-term service contracts in Iraq are similar to PSAs, the SC is signed without the agreement of the parliaments; it can lead to escalation between IOCs and future governments in Iraq. In order to have clean and transparent business for all participants, the IOCs should press to have the right regulation and a law to cover legal activities (investment), tax, and environmental issues related to the oil industry in Iraq. The political turbulence and mismanagement in the country make the development of the oil and gas industry and the country development questionable. There can be no concrete development of oil and gas fields without the parallel development of the infrastructure of the country; without this, turbulence may arise. The CAPEX for the exploration and development phases and OPEX are simulated with cost simulation taking into account the security factor. Even with high contingencies, still the 4.5-7.5¢/Mcf is low compared with that of other OPEC countries.

References

1. Senso Bujak et al. 1977; Hydrocarbon generation potential chart predicted from collide stratigraphical variations in kerogen types.

2. Thamer Khazaal Al-Ameri; Palynostratigraphy and the assessment of gas and oil generation and accumulations in the Lower Paleozoic, Western Iraq. Arab J. Geosci

3. Oil Ministry Feasability study

4. Tissot and Welte 1984, Hunt 1996; Akkas Field Stratigraphy.

**the paper published by the Oil Gas Journal (OGJ, 10/03/2011. Detailed analysis is available by the Author.*

3. Iraq's Ahdab oil field development limits contractor profitability

Ahdab is an oil field with a block area of 303 sq km located in the province of Wasit, 180 km southeast of Baghdad. The field was originally identified by 1970s vintage 2D seismic data. The first well on the structure is Al Ahdab-1, discovered oil accumulation in the Zubair formation. This was followed by further appraisal drilling in the early 1980s. A total of seven wells were drilled by the Iraq National Oil Company (INOC) on the Ahdab structure.

The field is being developed by CNPC in joint venture with the state-owned North Oil Company/Middle Land Oil Company under the terms of a Technical Service Contract (TSC) concluded in 2008 – which was the first awarded contract. The equity in the service contract is divided: 75% (CNPC) and 25% (North Oil Company/Middle Land Oil Company).

The Ahdab development is not only oil production but also connected to the new oil-fired power plant being built in Shanghai Heavy Industries. Construction started in 2008 and is expected to take 48 months to complete. The design capacity of the plant is 1,320 MW. a 940 million USD contract and author estimate that up to 50,000 bbl/d of Ahdab crude may supply this plant at full capacity. Using crude oil to produce electricity is a very expensive alternative and not healthy for the environment. Consuming 50.000 bbl crude oil per day (average 80 USD per bbl crude oil) is a 4,000,000 USD/d loss for the Iraqi economy to burn one plant for crude oil.

Geology

The Ahdab field is an elongate anticlinal structure, with two prominent domes oriented along a NW-SE axis. The gross structural closure measures 30 kilometers by 5 kilometers. Oil pay has been proven in the Late Cretaceous Khasib and Mishrif formations and the Hauterivian Zubair formation. Zubair formation reservoir lithologically comprises interbedded quartzitic sands, shales and siltstones which were deposited in a series of migrating and overlapping delta lobes. This has resulted in the development of a complex, heterogeneous reservoir framework, with highly variable net reservoir thicknesses and reservoir quality across the structure. The Zubair reservoir contains light oil, with a gravity of 34°API. Khasib formation is chalky limestones. It is moderate and occasionally a good reservoir quality, with permeability up to 120 mD. This reservoir contains moderately heavy oil, with a gravity of 23°API. Mishrif formation is limestones with shale of shallow marine origin. The porosity averages 14% and has permeability of up to 50 mD. The Mishrif formation is sealed by the Tanuma tight limestone reservoir which contains heavy oil, with a gravity of 14°API. The reservoir Sulphur percentage is between 1.9-2.4%. The reservoir depths are approximately 3500 m.

Plan of Development

Ahdab's oil reserves are P50 estimated at about 1 billion barrels of oil. However, Author have estimated recoverable reserves based on contract term production plateau target of 115,000-120,000 bbl/d gradually from 2012 and total plateau over the next 20 years. This is expected to be around 670 million barrels. The remaining oil is held as technical reserves. Associated gas reserves of around 750 Bscf are also considered as technical resources because there have been no plans announced for their commercial development, in which case the gas is likely to be flared and/or re-injected. CNPC started its appraisal program in 2009. A 3D seismic survey has been completed and workover operations have started, drilling-vertical/horizontal. The field lies along the left bank of the Tigris River, in the main cereal growing province of Wasit. The use of horizontal wells will reduce the surface impact of the proposed development. Facility, gathering lines, trunkline to the nearest point to the Iraqi strategic pipeline is around 140 kilometers from Ahdab. From Ministry of Oil sources the tie of the connection should be to the 42-inch Iraq's Strategic Pipeline.

Production Strategy

The reservoir development plan taken in consideration with the 1997 PSC set a number of reservoir targets: 25,000 bbl/d on start-up and 50,000 bbl/d within four years. Peak production of 90,000 b/d was expected within six years from the production start date. If the new contract should be consistent with the original plan of development, then the production should include only the Late Cretaceous Khasib, Mishrif and Mauddud formations. The design capacity of the production facilities under the 2008 contract is expected to be greater than 100,000 b/d or in two-step facility development. Authors have assumed crude output will peak at 120,000 b/d in ten years from the start of production (figure 1).

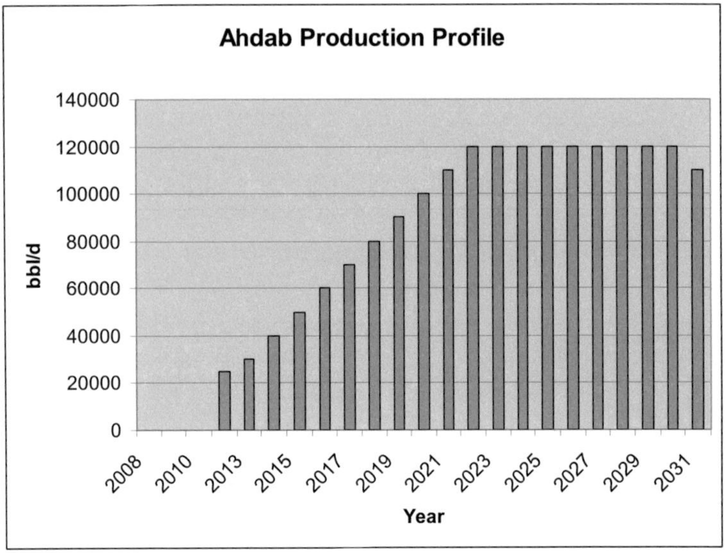

Figure 1: Ahdab Field Production Profile Target

Capital, and Operational Cost Expenditures

In 1997 CNPC estimated the cost of development at 660 million USD of which 350 million USD was allocated to drilling and completions. Required processing facilities, piping and storages was estimated at around 280 million USD. Operating Costs Expenditures (OPEX) were around 600 million USD over the life of the project, which equates to an average unit operating cost of 1.30USD/bbl.

The new costs (table 1) of the project have increased to a capital investment of 3 billion USD in 2009, which includes the mentioned power station (the station will be paid out of revenues generated from the sale of Ahdab crude oil), and an additional 1.8 billion USD in 2010. The total estimated CAPEX in 2009 and 2010 are around 3.9 billion USD for the development phase to produce a

120.000 bbl/d. The relationship total CAPEX to the total production is around 5.5 USD/bbl and OPEX average is around 5.2 USD/bbl after CNPC announced cost estimation. In table 2, Author calculations are according to the international market after CERA man power cost estimation. The CAPEX average around 2.9 USD/bbl and OPEX average around 2.6 USD/bbl. Even when Author adds 15% for contingency and 10% of escalation, which are very conservative numbers, we still have CAPEX of 3.6 USD/bbl and OPEX of 3.25 USD/bbl.

Year	bbl/d 0	Production Mn bbl	oss Reven Mn $	bbl/$	CAPEX Mn $	OPEX Mn $	Government Revenue Mn $	Government Net Revenue Mn $	Cost Recovery Fees Mn $	Remuneration Fee Mn $	Income Tax Mn $	Bonus Mn $	State Carry
2008	0	0,00	0		237	0	0	0,00	0	0	0	0	59,25
2009	0	0,00	0		385	0	0	0	0	0	0	0,5	96,25
2010	0	0,00	0		385	0	0	0	0	0	0	0,5	96,25
2011	0	0,00	0		392,7	0	0	0	0	0	0	0,5	98,175
2012	25	9,13	711,8	78,01	156,1	47,4	711,8	711,8	0	0	0	0,5	50,875
2013	30	10,95	793,1	72,43	159,2	57	396,55	396,55	396,5	0	0	0,5	-45,075
2014	40	14,60	1078,6	73,88	162,4	72,4	539,3	539,3	539,3	0	0	0,5	-76,125
2015	50	18,25	1375,2	75,35	165,6	88,4	687,6	687,6	687,6	0	0	0,5	108,4
2016	60	21,90	1683,3	76,86	168,9	105,1	841,65	841,65	841,6	0	0	0,5	141,9
2017	70	25,55	2003,1	78,40	172,3	122,3	1001,55	1001,55	0	589,80	206,43	0,5	73,65
2018	80	29,20	2335	79,97	175,8	140,4	1842,8	1842,8	316,2	176	61,6	0,5	0
2019	90	32,85	2679,4	81,56	179,3	159	2157,14	2157,14	338,3	183,96	64,386	0,5	0
2020	100	36,50	3036,7	83,20	182,9	178,2	2471,2	2471,2	361,1	204,4	71,54	0,5	0
2021	110	40,15	3407,1	84,86	124,3	195,7	2862,3	2862,3	320	224,84	78,694	0,5	0
2022	120	43,80	3791,2	86,56	63,4	212,6	3270	3270	276	245,28	85,848	0,5	0
2023	120	43,80	3867,1	88,29	64,7	216,8	3340,3	3340,3	281,5	245,28	85,848	0,5	0
2024	120	43,80	3944,4	90,05	66	221,1	3412	3412	287,1	245,28	85,848	0,5	0
2025	120	43,80	4023,3	91,86	67,3	225,6	3485,1	3485,1	292,9	245,28	85,848	0,5	0
2026	120	43,80	4103,7	93,69	68,6	230,1	3559,8	3559,8	298,7	245,28	85,848	0,5	0
2027	120	43,80	3504	80,00	70	234,7	3635,9	3635,9	304,7	245,28	85,848	0,5	0
2028	120	43,80	4269,5	97,48	71,4	239,4	3713,5	3713,5	310,8	245,28	85,848	0,5	0
2029	120	43,80	4354,9	99,43	72,8	244,2	3792,7	3792,7	317	245,28	85,848	0,5	0
2030	120	43,80	4442	101,42	74,3	249	3899,7	3899,7	323,3	219	76,65	0,5	0
2031	100	36,50	3775,7	103,44	0	223,1	3370,1	3370,1	223,1	182,5	63,875	0,5	0
2032	0	0,00	0		0	0	0						0
Totals		669,78	59179,1		3665	3462,5	48990,99	48.990,99	6715,7	3742,74	1309,959	11,5	

Table 1: Ahdab Field Technical Service Contract Conditions

Year	Oil Production mio bbl	Drilling/workover Workover mio $	Facilities mio $	Pipelines mio $	Total CAPEX	Operating Expenses Overhead- Workover Fixed Variable Total OPEX
2008	0,00	2	50	235	287	
2009	0,00	50	50	235	335	
2010	0,00	60	50	235	345	
2011	0,00	70	50	235	355	
2012	8,67	80	50	10	140	34,4
2013	10,40	90	10	10	110	41,6
2014	13,87	10	10	10	30	55,48
2015	17,33	10	10	10	30	69,32
2016	20,80	30			30	83,2
2017	24,27	30			30	97,08
2018	27,74	30			30	110,96
2019	31,20	30			30	124,8
2020	34,67	30			30	138,68
2021	38,14	30			30	152,56
2022	41,61	30			30	166,44
2023	41,61	30			30	166,44
2024	41,61	30			30	166,44
2025	41,61	30			30	166,44
2026	41,61	25			25	166,44
2027	41,61					
2028	41,61					
2029	41,61					
2030	41,61					
2031	34,67					
2032	34,67					
Total	670,92	697	280	980	1957	1740,28

Table 2: Ahdab Field Cost Estimation after CERA

Economics

Author estimates of the Ahdab field costs are based on Iraq's 2009 model Technical Service Contract (TSC) as illustrated in table 1. The key remunerative element of the TSC is a service fee comprised of petroleum costs and remuneration fees. Costs are recoverable on a dollar-for-dollar basis, subject to a service fee cap equal to 50% of petroleum revenues. The remuneration fee is payable to the contractor for the amount of oil produced, on a dollar-per-barrel basis. The maximum remuneration fee decreases according to the ratio of cumulative revenues to cumulative costs (R-factor). The R-factor is less than one; 6 USD/bbl; R-factor is 1-1.5 is 5.4 USD/bbl; R-factor is 1.5-2 is 4.9 USD/bbl and R-factor more than 2 is 3 USD/bbl.

The contractor is liable for corporate income tax at a rate of 35%. Net taxable profits are calculated as the annual aggregate of the remuneration fee actually received.

Author has used the economic assumptions for the cash flow model. The discount rate is 10% nominal. The discount date is 1/1/2010. Long-term inflation is assumed at 2% per annum from January 2010. Author has assumed that Ahdab crude will trade at less (2.5%) to the Brent price. The CERA price assumptions of Brent oil 81.28 USD/bbl in 2011, 87.85 USD/bbl in 2012, 91.69 USD/bbl in 2013, 94.94 USD/bbl in 2014, 98.86 USD/bbl in 2015, 102.56 USD/bbl in 2016, 104.44 USD/ in 2017. Wood Mackenzie's Brent price assumption is 77.63 USD/bbl in 2010, 84.50 USD/bbl in 2011, 80.00 USD/bbl in 2012, 74.28 USD/bbl in 2013, 75.77 USD/bbl in 2014 and 77.29 USD/bbl in 2015, escalating at 2.0% per annum thereafter. This equates to a long-term assumed oil price of 70 USD real (2010 terms).

The Internal Rate of Return (IRR) takes into account the timing of cash flows from Ahdab contract period. It is the discount rate which makes the net present value of a project equal to zero. The Ahdab annual discount rate (or interest rate) of 10% and for example a 24 year-long generation. In 24 years, a compounded 10% annual interest rate will multiply an original investment by a factor of 10, and an annual discount rate of 10% will decrease a later money receipt by a factor of 10. In table 3, we present the sum of Ahdab Net present Value (NPV) that results when the discounted value of the expected costs of Ahdab investment are deducted from the discounted value of the expected returns.

Year	Field Cash Flow $ Mn	NPV $ Mn	Company Cash Flow $ Mn	NPV $ Mn
2008	-237		-237	
2009	-386		-386	
2010	-386	-290	-386	-290
2011	-393	-269	-393	-269
2012	-204	-127	-204	-127
2013	180	101	180	101
2014	304	156	304	156
2015	433	202	433	202
2016	567	241	567	241
2017	500	193	404	156
2018	114	40	85	30
2019	119	38	89	28
2020	132	38	99	29
2021	146	38	109	29
2022	159	38	119	29
2023	159	35	119	26
2024	159	31	119	24
2025	159	29	119	21
2026	159	26	119	19
2027	159	24	119	18
2028	159	21	119	16
2029	159	20	119	15
2030	142	16	106	12
2031	118	12	88	9
Total	2421	614	1813	475

Table 3: Ahdab Net Present Value

Profit oil is the net revenue/production after tax, bonus etc. have been deducted. It is divided between the International Oil Company (IOC) and Host Government (HG). The portion of the profit oil allocated to the IOC is usually negotiable and stipulated in the agreement. The basis of the spilt varies in different countries and may differ even within the same country depending on prospectivity, the fiscal regime of the HG and also the level of production which could be a flat rate or sliding scale. Figure 2 shows the split between the CNPC, costs and the HG. In this article Author are not showing the cumulative net cash flow in discounted rate of 10% curve. But Author mentioned above what means the discount rate of 10% for the example of 24 years. The rough price sensitivities analysis for the remaining present value calculation after tax between (-20% to +20%); gradually the production will increase, as will the oil price, CAPEX will decrease and a slight decrease or consistency for the OPEX. This means a remaining present value after tax of around 650 $Mn to 1100 $ Mn.

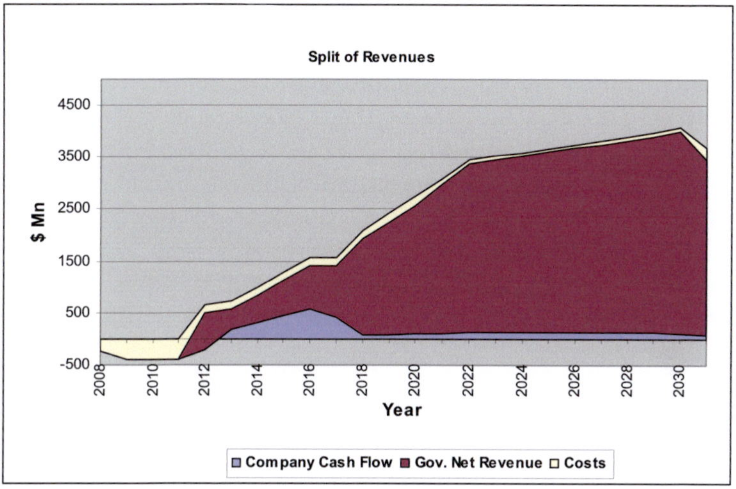

Figure 2: Ahdab Field Split of Revenue

Conclusion

It is clear that there are differences in the costs estimation between our costs which are based on CERA cost estimation and the CNPC cost estimation. Even when Author adds 15% for contingency and 10% of escalation, which are very conservative numbers, we have CAPEX of 3.6 USD/bbl and OPEX of 3.25 USD/bbl. The NPV of the CNPC is very high compared with the IOC's worldwide figures under the TSC and or even PSC terms. The IRR of 10% of the TSC is reasonable. If we isolate the remuneration provision that we have just briefly explained, we can say that the economic return to the contractor is modest in comparison to any other type of contract used in the oil industry. But if we look at the broader picture, we can notice that the prize may be great. The contract remuneration in Iraq's contracts is a matter of scale.

Using crude oil to produce electricity is a very expensive alternative and not healthy for the environment. Consuming 50,000 bbl crude oil per day per average 80 USD per bbl crude oil is a 4000.000 USD/d loss for the Iraqi economy from burning one plant for crude oil. Even if fuel oil is partially used, this is not the best alternative. Available gas cap and or associated gas can be used for a small power plant instead to flair it and or inject it again. In case the transport of the gas to the main pipeline not economic, there is the option in the same central plant facility to treat the gas and produce it for the domestic and industrial use.

Geologically the Khasib is chalky limestone, the Mishrif is limestone with shale and the Zubair is quartzitic sands, shales and siltstones formations. They are all virgin layers with different compositions, different crud classifications, producing single or comingled, the control of gas overrunning and possible future water under running, sand problem and H2S etc. Therefore, it is recommended to plan different scenarios of development with the best technical simulate.

In order to optimize reservoir management in this field, the Federal Ministry of Oil (MoO) should describe the Maximum Efficiency Rate (MER) for the wells. It is efficient for the reservoir, wells and the stability of the field plateau.

After Iraqi sources, there are huge shortages in all Iraqi expertise disciplines in the MoO and national oil companies which are very important for the audit and the development phase of the field.

**the paper published by the Oil and Gas Journal (OGJ in 01.08.2011). Detailed analysis is available by the Author.*

4. Hydrocarbon reservoir potential estimated for Iraq bid round blocks

Introduction

The 4th bid round was announced in May 2011 and the bidding is to be held on April 2012, this is the first round to offer exploration contracts rather than the technical contracts offered in the previous three rounds. Originally 46 companies qualified for this round but this is split into operating companies and non operating companies who can compete on contracts but only as a partner led by an operating company. Twelve areas are on offer covering 81,700 square kilometres with the blocks having a mix of oil and gas productivity (Table 1). However, many of the blocks contain oil and gas shows and some of them very close to previously discovered oil and gas fields. Eight of the 12 blocks are situated in the West of the country along the Iraq-Syria-Jordan, and Iraq-Saudi borders (Figure 1) /2/.

The Western and South western desert is estimated to have high potential for hydrocarbon, where many thousand of kilometres of 2D seismic data were acquired and the own state Oil Exploration Company (OEC) has discovered several fields. The exploration of oil and gas has been decreased during Iraq's wars and sanctions. The most of the more than 60 exploration wells were drilled in stable and peaceful time of Iraq, with discovery success of more than 70%. The author estimate that at least for the first exploration phase 250 to 300 exploration and appraisal wells need to be drilled in the existing 80 appraised and partially appraised fields.

The Western Desert Exploration Area was divided into nine contract areas in the 1990s. The blocks cover a large area of Western Iraq - extending from the Euphrates River in the East, to the Jordanian border in the West, and the Saudi Arabian border in the South. The exploration period for Western Desert acreage was split into two periods of 3 years and 2 years respectively. Under Phase 1, contractors were obligated to reprocess existing 2D seismic data, acquire new 2D data, and drill two exploration wells. Phase 2 commitments included the acquisition and processing of 3D data, and two additional exploration wells.

In the previous blocks distribution the Indian OVL has operated in the block (8)* in the Western and South western of Iraq and drilled 7 exploration wells. The OVL has estimated the block 8 include the Abu Khaima field has the potential to produce up to 300.000bbl/d. The MoO has renegotiated the previous contract under the new exploration service contract and the possible again renegotiation with the proposed petroleum law.

In the Northern part of Iraq provinces are Arbil, Sulaymania and Dhouk which today called (Iraqi Kurdistan), the given exploration blocks are still point of not solved tension between the central government of Iraq and Kurdistan Regional Government (KRG) /1/.

Currently Ministry of Oil MoO promoted areas under new contract terms to attract foreign oil company investment. The OEC have mapped twelve exploration blocks with areas between less than 10000 to more than 14000 square kilometres. The current announced exploration development blocks, two to the Syrian border, three to Jordanian-Syrian border, three to the Saudi Arabian border, and two to the North east and South east on the Iranian border and two in the middle South of Iraq.

The mentioned regions are thought to be highly prospective and contain numerous undrilled prospects. Only four exploration wells have been drilled on the blocks and the seismic coverage is old and of poor quality. The OEC had been planned to drill more than

300 exploration and appraisal wells in the Western and South western region. High unlikely the Ministry of Oil (MoO) and their own state companies the poor management states and shortage of technical knowledge that this programmes was achievable /1, 9/.

In the W-NW-SW of Iraq contains five oil discoveries (Abu Khaima, Ekhaidar, Khleisia, Salman and Samawa), two gas and condensate discoveries (Akkas and Diwan) and 300 prospects and wide number of upsides in the North East-west and Southwest of the Iraqi desert.

The geology and geophysics (G&G) survey are covered part of Iraq area. Most of the results of the seismic are poor which include Western and South western to the South of the country. To the North East region of the Northern folded platform more than 50 large, many smaller fields and more than 300 possible prospects and upsides to the North East-south west trend. Remaining are many fields to be discovered in the high exploration potential areas of the Mesopotamian Basin and Northern Folded zone. The OEC has discovered more than 110 prospects in the Western and South Western desert. The potential of discovering high potential reservoir is very considerable which may change the Iraqi total reserves to 300 or more billion barrel of oil and 500 trillion bscf of gas.

The objective of this article is to identify and evaluate the petroleum systems of prospects and leads, calculate the reserves and evaluate the hydrocarbon potential and economics of the 12 blocks using existing literature and available data published by the Ministry of Oil data.

Block Name	Province	Area Km2	Expected	Additional
1	Nainawa (Nineva)	7300	gas	oil/condensate
2	Anbar, Nainawa (Nineva)	8000	gas	oil
3	Anbar	7000	gas	condensate
4	Anbar	7000	gas	condensate
5	Anbar	8000	gas	condensate
6	Najaf, Anbar	9000	gas	condensate
7	Qadisia, Babel (Babylon), Najaf, Muthana (Samawa)	6000	oil	associated gas
8	Diyala, Wasit (Kut)	6000	gas	condensate/oil
9	Basrah	900	oil	associated gas
10	Muthana (Samawa), Thiqar (Nasiriya)	5500	oil	associated gas
11	Najaf, Muthana (Samawa)	8000	oil	associated gas
12	Najaf, Muthana (Samawa)	8000	oil	associated gas

Table 1: Exploration development blocks with a mix of oil and gas productivity

Leads were by Author and OEC identified using the seismic data available over each Block. The seismic data was tied into wells on or near the blocks. Geology and Petrophysical analyses of wells, fields and neighbour fields/wells were evaluated. Reserves were then estimated for each lead.

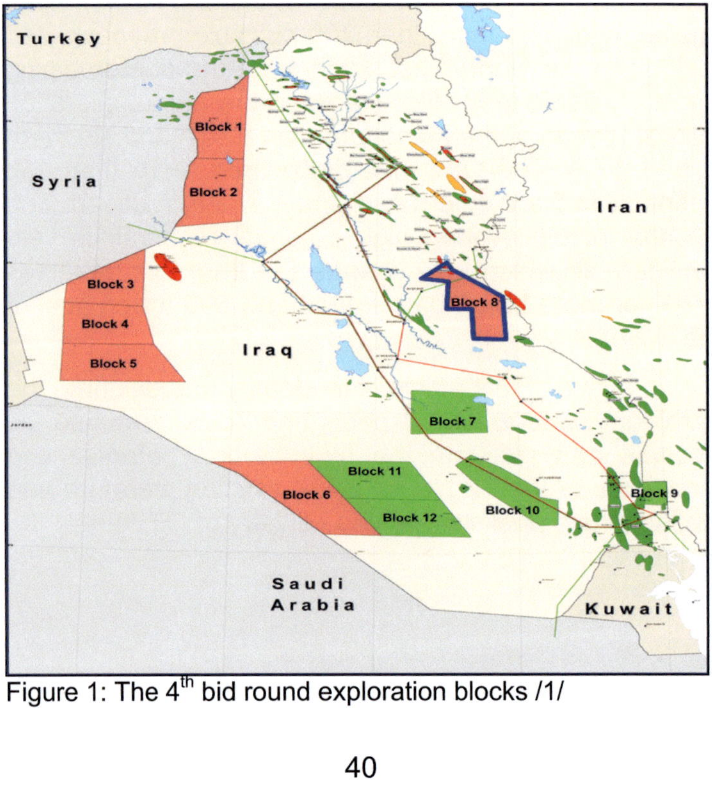

Figure 1: The 4th bid round exploration blocks /1/

Estimated Exploration Potential

Iraq's oil reserves are estimated at about 110 billion barrels of oil. This figure includes commercial oil reserves of about 102 billion barrels. It also includes technical reserves of about 8 billion barrels, which are not under development. In October 2010, the MoO published reserves data for 66 oil fields with combined reserves of 143 billion barrels. Discoveries in Kurdistan were excluded from this list. The reserves of many of the fields had not changed from previous Iraqi figures. The largest increases were made at West Qurna, which jumped from 21 billion to 43 billion barrels, and Zubair, which increased from 4.1 billion to 7.8 billion barrels. By some Iraqi estimates, there is the potential for 350 billion barrels to ultimately be discovered in Iraq. This view is based on the fact that some 400 structural anomalies and leads have been mapped in Iraq but never drilled - more than half of which are considered to have a high degree of certainty. Reserves growth is expected in three major geographic provinces; Mesopotamian sedimentary basin, Western Desert and Kurdistan region.

From the available geology-geophysics survey, well tests and petrophysics input using analogue fields to correlate the same structures, the Author make the first phase conservative reserves estimation as following:

Area	Estimated Oil Reserves Billion bbl
Northern Fold	34.2
Mesopotamian Basin	68.3
West and South West Desert	76.3

Table 2: Estimated oil reserves per area

Author reserves figure does not include any of this yet-to-find potential and therefore reflects a fairly conservative view.

Iraq has estimated gas reserves of between 100 and 110 tcf of gas, around 70% of which is solution gas, primarily associated with the giant fields in the southeast of the country. Reserves of around 15 to 20 tcf of non-associated gas are contained within seven fields, mainly in Northern Iraq. As with oil, the potential for growth in Iraq's gas reserves is very high, given the limited extent of exploration. Furthermore, exploration has been focused on oil targets, which has meant that many gas discoveries have never been fully appraised. The discovery of the Akkas gas field in Ordovician sandstone indicated the Western Desert as a potential gas province. Iraq's yet-to-find gas potential has been estimated at about 500 tcf - split 60/40 between non-associated and associated reserves.

Geology and Exploration Potential

The oil reserves are found in two distinct basins in Iraq: the Zagros Basin and the Arabian Basin. The Arabian Basin is subdivided into several separate sub-basins: the Zagros Mesopotamian Basin, Widyan Basin, Sinjar sub Basin and the Rutbah uplift. Iraq has a very limited coastline (i.e. without the claimed zone of at least 100

Km from the existing border with Kuwait) and all of its oil fields are located onshore.

The majority of fields and discoveries are in the Folded Zone and the Mesopotamian Basin, which has the fields in Northern, East northern and Southern of Iraq. The majority of Iraq's hydrocarbon reserves are in Cretaceous and Tertiary reservoirs. Whilst potential still remains in the underexplored Jurssic, Triassic and Palaeozoic systems. The not explored areas offer a diverse range of opportunities. The relatively unexplored blocks are; block 1, is located in Khleisia high and Sanjar trough, two exploration wells were tested from Miocene and oil shows in Misozoic. Block 2, one exploration well with oil stains in Cretaceous, block 3, near Akkas field, block 4, near Akkas and Risha field in Jordan, block 5, near Akkas and Risha field, block 6, no exploration wells, a number of structure can be discover. The last three blocks could be find Palaeozoic, Ordovician and Jurassic targets. Block 11, is in the Salman zone, one exploration well, block 12, one exploration well, both blocks could be Cretaceous and Jurassic. Block 8 located on the border between Mesopotamian Basin and the Folded Zone, and block 9, near one exploration well in Mesopotamian Basin, both of them lie within well explored and recognised petroleum zones. A block 7 is located in the petroleum rich Mesopotamian Basin, near one exploration well and close to existing appraised discoveries, possible target is Cretaceous. Block 10 on the western boundary of the petroleum rich Mesopotamian Basin, located on border between Mesopotamian Basin and largely unexplored Salman Zone, possible discoveries is Cretaceous /1, 6/.

The Unstable Shelf in the north east of the country can be further subdivided into three distinct tectono-physiographic zones;

- The thrust or Zagros zone; a narrow belt stretching NW-SE along the north eastern borders with Iran and Turkey, containing highly deformed Palaeozoic and Mesozoic rocks.
- The high folded zone; formed by Zagros deformation during the late Cenozoic, this zone is characterised by long, asymmetric anticlines separated by narrow synclines and folds are often eroded to expose Mesozoic rocks at outcrop.
- The low folded zone or foothill zone; this zone forms the south west section of the unstable shelf and can be further sub divided into the Kirkuk-Hamrin and Mosul-Butmah subzones. Anticlines of varying lengths and heights are located throughout the area and towards the south west boundary, the folds generally flatten.

The Stable Shelf occupies most of the west, central and east/south eastern part of the country and can be subdivided into three tectono-physiographic zones or basins;

- The Mesopotamian zone or basin; this is a large area of flat lying topography stretching from northern central Iraq (where it forms part of the boundary with the unstable shelf) south to the Kuwait border. There is a regional dip to the east/north-east and the western margin is believed to be bounded by NNW-SSE fault zones The subsurface is characterised by relatively gentle anticlines and horst structures buried beneath undeformed to gently deformed Neogene formations, which is overlain by fluvial Quaternary deposits.
- The Salman Zone; this zone runs north to south down the western edge of the Mesopotamian basin, a remnant Precambrian palaeohigh that was reactivated during the late Carboniferous. Basement depth in the Salman Zone shows a deepening from 5 km in the west to 8-9 km along the eastern boundary.
- The Rutbah-Jezira Zone covers the vast Western Desert of Iraq, is an inverted Palaeozoic basin and in the north is bounded by the southern margin of the Unstable Shelf, as marked by large anticlines such as Jebel Sinjar. It is also cut by the east-west trending Anah Graben, which subdivides it into the southern Rutbah zone and the northern Jezirah zone. The Rutbah zone is dominated by the Rutbah uplift and the Ga'ara depression, whilst the Jezirah zone is dominated by the Khlesia high.

The principal tectonic elements are shown together with the outline of the evaluated blocks on Figure 2.

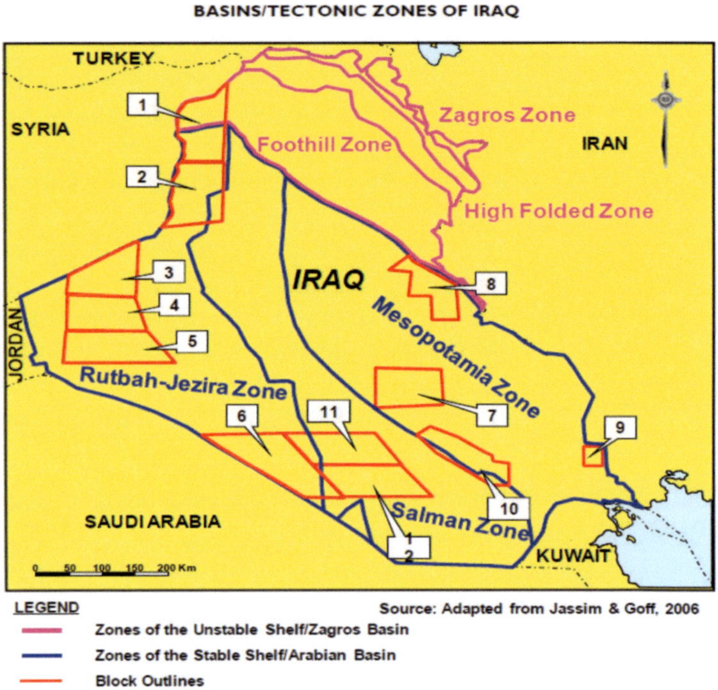

Figure 2: The main tectonic elements of Iraq /1, 4, 5/

There are a large number of structural elements (lineaments, structures and faults) present in Iraq. These features have influenced depositional patterns and facies development and in some instances exert significant influence on the development of the aforementioned tectonic zones. There are three major Precambrian fault systems identified in Iraq, and these are;

- The NW-SE trending "Najd Fault System"
- The NE-SW trending "Transversal Fault System"
- The N-S trending "Nabitah Fault System"

The three fault systems are the cause of the three major paleostructural trends observed in Iraq and were re-activated at various times during the Phanerozoic. Of the NW-SE trending faults, the Euphrates Boundary Fault controls the margin of the Quaternary Mesopatanian Basin, the Ramadi - Musaiyib Fault controls the location of the subsurface West Baghdad structure, the Tikrit – Amara Fault system controls the location of subsurface strike-slip related Late Cretaceous structures, including the East Baghdad oilfield and the Makhul – Harmin lineament comprises a line of anomalously high surface anticlines defining the SW limit of the folded zone. Also oriented NW-SE is the Qamchaqua Lineament, which controls the edge of the Qamchaqua carbonate shelf in Kurdistan /1, 4, 5, 6, 7/.

The NE-SW trending "Transversal Fault System" includes the western portions of the Sinjar – Herki and Kouch – Dohuk faults, which probably correspond to the southern and western bounding faults of the Sinjar Trough. The location of the Anah Graben is controlled by the Anah – Qalat Dizeh Fault Zone. Regional anticlinoriums are formed by the Rutbah High and the Khelisia High, which were initiated in the Late Carboniferous and represents an inverted Paleozoic basin. The Bekkne – Hadar Lineament forms the SE boundary of the Moussul High. In northern Iraq, a range of NW-SE oriented features may have controlled facies patterns and influenced the response to Tertiary deformation, for example the Amij – Samarra - Halabaj fault apparently controls the strike of the Quamchaqua Formation of Albian – Barremian age. East-west structural lineaments are less common than those oriented NE-SW or NW-SE. Examples of east-west structural features include the Sinjar Trough, the Tayarat Graben and the Anah Graben. East-west structural lineaments may function as transfer zones between NE-SW and NW-SE lineaments. Some east-west structures, for example the Anah Graben cross-cuts paleohighs, with the Anah Graben separating the Rutba High and the Khelisia High. Paleostructural trends oriented N-S include the Salman Zone in southern and central Iraq. The Salman Zone is bounded by the Abir Jir Linearment to the east, which separates stable and unstable shelf. The structures of the Rumalia, Zubair and Ratawi Fields in SE Iraq are north-south oriented highs, which formed by fault re-activation and salt movement during the Late Cretaceous, Figure 3 /4/.

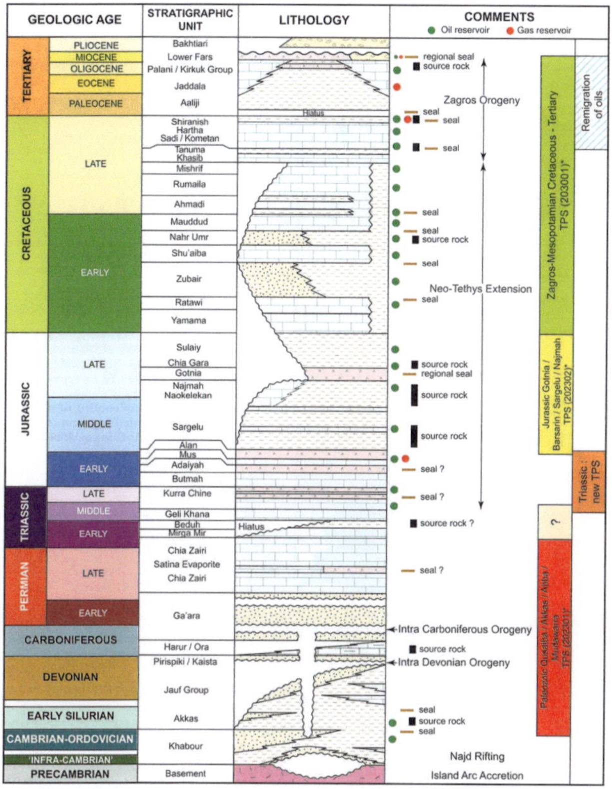

Stratigraphic distribution of source, reservoir and seal lithologies in Iraq

Figure 3: Total petroleum systems according to Verma et al. 2004 /4, 5, 7/

Seismic Interpretation

There are five leads in block 1 identified. Upper Cretaceous Shiranish and a number of potential Cretaceous, Jurassic and Triassic Kurra Chine and Ordovician Khabour levels. Six leads have been recognised in block 2, in the Triassic Kurra Chine and the Ordovician Khabour levels. Blocks 3, 4 and 5 are located in western Iraq in the Jezira-Rutbah zone; six surveys were attained from 1979 to 1990 with processing datum's varying from 300m to 700m above sea level. The seismic data was acquired with Vibroseis and is mainly 24 fold. Data quality is mainly poor, although the shallow horizons are often degraded; this shows that the deeper horizon can be partially identified. Block 3 has eight leads. These are defined by the Gaara and Khabour depth structure. Block 7 is located in central Iraq in the Mesopotamian Basin. Data quality is fair to poor; there is a significant area in the west of the block that has no seismic coverage available during this work. Overall the potential structures in Block 7 are relatively small and a total of ten leads were identified. Block 8 is located in eastern Iraq, close to the Iranian border. The seismic data was acquired with dynamite and is 24, 12 or 6 fold. Data quality is fair.

There are no wells situated in Block 8, however, there are a number of wells associated with fields and discoveries surrounding the block. Mansuriyah-1 (1978), Jeria Pica-1 (1975), Nau Doman-1 (1978), Naft Khaneh-1 (1920) and Tel Ghazal-1 (1980) are all situated to the North east of the Northern part of the block targeting large anticlinal structures. Hydrocarbons were discovered primarily in the Miocene age Jeribe Formation. Badrah-1 is located east of the southern part of the block and was drilled in 1978 targeting a large anticlinal structure. 32 API oil was discovered in the mid Cretaceous Maududd Formation. Six leads have been identified at Upper Fars and Jeribe-Euphrates level. It is important to note that in Block 8, with improved seismic coverage and/or more time to carry out interpretation, it is possible that the identified leads may increase in size and also that more leads may be identified throughout the block. The leads have been modelled at Upper Fars and Jeribe-Euphrates level, but in some of the larger structures there may be the possibility of closure into the Cretaceous section. Block 9 is located south eastern Iraq adjacent to Iran in the Mesopotamia Zone. Two seismic surveys have been acquired over the Block in 1973 and 1978 by Petrobras. The seismic data were acquired with a dynamite source and is 6 and 12 fold. Data quality is fair to poor. There are no wells located within Block 9. Locations of five wells located outside the Block were available although only formation tops for two wells (Nahr Umr-2 and Nahr Umr-3) were available. The Nahr Umr Field is located 5 kilometres to the west of the Block and is intersected by seismic data. The Sindabad Field is located 1 kilometres to the south of the field and is also intersected by seismic data. The Zubair, Majnoon and Siba Fields are also located nearby and on trend with the Block.

Three Leads have been identified. There is potential for a large number of reservoirs to be hydrocarbon bearing as the structure is expressed throughout the overlying horizons. The nearby Majnoon Field has up to twelve oil bearing reservoirs down to the Cretaceous level. Three leads have been identified at Upper Zubair level, which was oil bearing in the Luhais-1 well. All three structures are very subtle and do not show closure on the depth maps. Blocks 6, 11 and 12 are located in south-western Iraq. Data quality is fair. Safawi-1 is located 15 kilometres south of Block 6 but on a seismic line which extends into the block. Ghalaisan-1 is located in Block 11 and was drilled in 1960 to a depth of 1991m as a stratigraphic test on a positive gravity anomaly. A total of nine leads were identified in Block 6 at the three different stratigraphic. Samawa-1 is located approximately 50kilometres to the east of Block 11 and was drilled in 1958 to a depth of 3841m to explore the Samawa prospect to Jurassic level. Drilled in 1988 to a depth of

5475m, 40 API oil was discovered in the Ratawi Formation. Ubaid-1 is located 20 kilometres south of Block 12 and was drilled in 1960. Leads were identified at all potential reservoir levels and in a variety of different trap types. There was also a large distribution in the size of the leads. Shawiya-1 is located in Block 12 and was drilled in 1960 to a depth of 1784m as an exploratory well located on a positive gravity anomaly and on the south west flank of a surface fold. Three leads have been identified at Upper Zubair level, which was oil bearing in the Luhais-1 well. All three structures are very subtle and do not show closure on the depth maps. Therefore areas of potential closure have been identified visually on the maps and identified as leads /1, 8, 9, 10, 11, 12, 13, 14/.

Petrophysical Analysis

Most of the wells were drilled in the 1960s or 1970s and have antediluvian log consecution, commonly consisting of only an SP and a resistivity log. The problem in many wells is the determination of a good porosity, water salinity, mud resistivity and formation mud resistivity. With porosity determination is only possible in six exploration wells; water salinity was only available for three wells. The abstraction of mud resistivity does eliminate determination of Rw. The water saturations determined are therefore estimates /1, 9/.

Reservoir Properties and Well Testing

Collected data from producing and non producing analogue fields and this include formation properties and their correlation to similar formation in the mentioned blocks area, and the average of the formation volume factors of neighbour fields have been taken. Analogue data for fields in Iran, Syria, and Saudi Arabia have been taken from the published data. Information on potential reservoirs was taken from Oil Ministry Reports on northern and western Iraq, together with Aqrawi et al. (2010)[1] and Jassim and Goff (2006) /4, 7/.

Ten principal reservoir intervals are; the Khabour Formation (Cambrian-Ordovician), the Kurra Chine Formation (Late Triassic), the Najmah Formation (Late Jurassic), the Yamama Formation (Early Cretaceous), the Zubair Formation (Early Cretaceous), the Nahr Umr Formation (Early Cretaceous), the Mishrif Formation (Late Cretaceous), the Hartha Formation (Late Cretaceous), the Jeribe Formation (Early Miocene) and the Euphrates Formation (Middle Miocene). Available porosity and permeability data has also been summarised on the reservoir quality Oil Ministry maps /1, 4, 5/.

The Tel Hajar-1 well in block 1 was tested in 1979 and produced 100 bbls/day of light gassy oil from an 11m interval in the Kurra Chine and 300 bbls/day of light gassy oil from a 40m interval in the Serikagni. Sargelu and other potential formation should be retested again for ability of hydrocarbon contains. In the same block the Abtakh-1, should be tested in the Far, Jeribe, Shiranish/Hartha, Mushorah, Bitumah, Kurra Chine and Geli Khan Formations. In block 2 – Khleissia-1, should be tested the Jeribe, Euphrates/Sarikagni and other potential formation of the Ordovician. Block 11 – well Ghalaisan-1, which contain Dammam, Hartha, Kifl, Rumaila, Nahr Umar, Zubair, Ratawi, Yamama, Sulaiy, Gotina and Najmah should be tested. In block 12 – well Shawiya-1, the potential formation are Kifl, Rumaila, Nahr Umar and possible other formation, Author suggestion to drill deeper than already drilled.

The Akkas Field was used as an analogue field for blocks 1, 2, 3, 4, 5 and 6 of which the Khabour and Ga'ara are the main gas reservoirs. The Kurra Chine is the next most important formation which generally contains oil. The Khabour formation is also the main reservoir in Leads block 6 and block 11 /13, 14, 11, 7/

Akkas and Khabour Formations

The Akkas Field is located in the Northwest region of Iraq near the Iraqi–Syrian border. The Akkas-1 well produced gas condensate, wet and dry gas. The specific gravity of the tested gas in Akkas-1 was between (0.726 – 0.6953), and for the gas condensates was 0.7792. There are also indications of light oil (density 0.8326 gm/cm3). An Akkas-4 test showed natural gas; however we believe that beside the gas, condensates may also exist /14/.

The Khabour Formation is the oldest known sedimentary rock unit in Iraq. It consists entirely of siliciclastic sequences, comprising thin-bedded, fine-grained sandstone, quartzite graywackes, and silty micaceous shale with some intercalation of dolomite and limestone. The base of the formation has not been reached in wells, and it may be more than 1,000m thick; it was deposited in shallow marine inner to outer neritic environments, which prevailed over the entire eastern part of the Arabian Plate. In the Akkas Field, the Khabour Formation is found at depths of below 2,310m and exhibits good reservoir properties. It comprises sandstone with silty shales and has a gross thickness of approximately 38m and an average thickness of 25m. Core analysis of the reservoir intervals generally indicated an average porosity of 10% and permeability of 500 mD. Its depth in Akkas-1 is at 2,040m below sea level. Formation temperature is 146°C at 3,300 m. The high temperature is caused by the Silurian shales' radioactivity and decreases downward in the Khabour Formation to normal gradients. The gross hydrocarbon reservoir column is about 80 m.

A secondary reservoir is provided by the Akkas Formation found above the Khabour Formation at a drilled depth of 1,465–2,326 m. It is comprised of a sandstone unit inter-bedded within the basal hot shale unit, with a gross thickness of approximately 10m and net thickness of 1.5m. Core analysis of the reservoir intervals indicated porosity of 17% and a permeability of 500 mD. Light oil (42° API) is present only in the thin sandstone horizon sandwiched between the hot shales at the base. It is therefore considered a minor reservoir. The overlying Silurian shale of the Akkas Formation is an effective local regional seal.

Ga'ara Formation

Can be divided into Lower and Upper units, similar to its time equivalents in Saudi Arabia (A-Khlata and Unayzah formation) (Al-Laboun, 1988; Aqrawi 1998). Its thickness is at least 120m at Tlul Afyef and Wadi Duwaikhal, and 450m in well Kh-5/1. It is also present in shallow boreholes and trenches in the western desert (Jassim 2006c). In west Kifl-1, a 50m thick section of the Ga'ara formation consists of cross-bedded sandstones.

In wells Atshan-1, and west Kifl-1, sandstones and shales lie beneath the carbonates and high GR API shales of the Chia Zairi and upper most Ga'arra formations. Thicknesses vary from 83m in well Kh-5/2 in western Iraq, to 442m in well Atshan-1, 514m in well west Kifl-1, and 631m in well Jebel Kandl-1. Wells Mityaha-1 and Diwan-1 partially penetrate this unit. A well at Burgan in Kuwait penetrated a thickness of 586m (Khan, 1989). The super sequence is absent in wells Akkas-1, Qaim-1 and Khleisia-1 and in much deeper west desert keyhole wells (Kaddouri, 1996, Aqrawi 1998, Al-Hadidi, 2007). Rocks in this super sequence are assigned to the upper part of Ga'ara formation.

Kurra Chine

Triassic reservoir rocks in Iraq contain <1% of the countries proven reserves (Al-Sakini, 1992). Although several fields in NW Iraq and NE Syria have produced significant volumes of oil and gas from Triassic reservoirs during production tests. The most important reservoir rocks occur in 500 m thick Upper Triassic Kurra Chine formation, for example in the Butmah, Sufaiyah and Alan oilfields in north Iraq, and in the Kurra Chine and Butmah formation in the Suedie, Rumelan, Jebissa and Tishreen fields in Syria. The main producing interval in Iraq lies in the carbonates of the upper part of the Kurra Chine formation, which occur between two thick anhydrite intervals.

Most successful tests of the formation in Iraq are reported be from carbonates in the upper part of the Kurre Chine. At the Alan field, located north of Mosul, reservoirs in Kurra Chine formation are composed of limestones and anhydrites apparently with little porosity and production is probably controlled by fractures. Free oil was encountered at depths of 2569-2570 m in Alan-1 and gas shows at 2564m. In the Alan 2 well, productive intervals in the Kurra Chine formation consist of fractured anhydritic, dolomitic limestones with intergranular porosity. The interval 2408-2466m produced an average of 343b/d of 29.6 API oil. Gas and condensate came from 2762-2799m.

Well Ibrahim-1 found porous dolomite in the lower part of Kurra Chine formation at 2756-2778m and fractures were recorded in a cored dolomite interval at 3078-3165m. Reservoir evaluation suggested a pay which was approximately 82 m thick (Sadooni, 1995). Ibrahim-1 produced small volumes of 26 API oil and gas from the lower part of Kurra Chine formation, and tar and dead oil at 2756-2778m. Oil and gas was observed in the fractured dolomite interval (Sadooni and Alsharhan 2004).

In the Sufaiya field, well Sf-1 tested gas and condensate at 2884-2906m, and oil and water from 2927 to 3045m (Sadooni, 1995, Sadooni and ALsharhan, 2004).

The Kurra Chine formation at Butmah contains 35 API oil (Dunnigton, 1967). Gas bearing zones were encountered in Butmah-2 at 3198-3233 m and 3253-3255m, and an oil zone at 3276-3278m.

Oil and gas shows were reported from by Sadooni (1985) from wells Atshan-1, Jebel Kandl-1, and Kifl-3; oil shows were reported from wells Qailan-1, and Tel Hajar-1, and heavy oil shows were reported from well Mileh Tharthar-1.

Golf Keystone (2009) reported an oil discovery in the upper part of the Triassic in well Shaikan-1, about 50 Km north of Mosul. Multiple drilling occurred in 81 m gross interval at 2402 to 2483m. These probably correspond to thin porous beds in the Kurra Chine formation. They flowed 200bbl/d of 40 API oil with GOR of approximately 1000scf/bbl.

In the Euphrates Graben, oil was tested from both leached dolomites and sandstones in the lower Mulussa formation (Al-Sakini, 1992; de Ruiter et al., 1995; Litak et al., 1998). At the Rumelan field, 48 API oil is present in Kurra Chine dolomite (Dunnington, 1967).

Yamama Formation

The Yamama formation contains hydrocarbons at 26 structures in southern Iraq including the West Qurna, North Rumaila and Majnoon fields (Jassim and Al-Gailani, 2006). The Formation consists largely of shallow water carbonates. Reservoir limestones were deposited in oolitic shoals as low energy, shallow marine carbonates and ramp / slope facies in central Iraq. The average thickness of the Yamama reservoir is 240m. Reservoir quality is best in central Iraq and is poor in eastern Iraq, average porosities are 6-12% and permeabilities range from tens to hundreds of milidarcies. Porosity is dominated by intergranular macroporosity and some intragranular to chalky matrix microporosity, with local secondary porosity. Such micro-porosity systems often have high water saturations if water-wet, or are bitumen-plugged if oil wet. At the Rumaila, West Qurna structure, the reservoir potential of the Yamama formation increase from South to the North. Flow rates are very variable; a maximum test flow rate of 19,000bbl/d was recorded in well Jerishan-1 (Sadooni-1, 1993b).

Zubair Formation

The most important sandstone reservoir in southern Iraq is the Zubair formation with 30% of Iraq hydrocarbon reserves, recorded as oil bearing in 30 structures including Rumaila North and South and Zubair (Al-Naqib 1967, Jassim and Al-Gailani, 2006), as well as East Baghdad in central Iraq. The Zubair forms a secondary reservoir at Ratawi, Tuba and Luhais, and contains smaller reserves at Majnoon, Halfaiya and Huwaiza (Jassim and AL-Gailani, 2006). The East Baghdad and Balad fields and Mileh Tharthar-1 well lie just within the sandstone facies (Ali and Aziz, 1983, Al-Ameri and Batten, 1997). Sandstones of the Zubair Formation were deposited in a delta / pro delta to inner shelf setting and consist of interbedded sandstone and shale 300-400m thick. The Formation grades north westwards into shales of the Balambo Formation. The proportion of sand increases to the Southwest, where the more proximal facies are developed and the formation is thickest in Southern Iraq. Reservoir potential is greatest in the southern part of Central Iraq where coarse grained, shallow marine clastics are developed. Net to gross in Zubair formation is around zero near the Iranian border; a thickness of over 200m of sandstone is present in the Zubair field and the Karbala area of cenral Iraq. Porosities likewise increase from an average of 15% near the border with Iran to around 30% in wells in Salman zone such as Kifl-1(Al-Naqib, 1967, Jassim and Goff, 2006a).

Mishrif Formation

The Mishrif is the most important carbonate reservoir in SE Iraq (Aqrawi et al., 1998) and contains oil at 32 structures (Jassim and Al-Gailani, 2006). The largest accumulations are in the Rumaila North, Rumaila South, West Qurna, Zubair, Majnoon and Halfaiya fields, located on large scale North-south trending anticlines. At least 15 other commercial oil accumulations in the Mishrif formation have been discovered in SE Iraq; Abu Ghraib, Ahdab, Amara, Buzurgan, Dujaila, Jebel Fauqi, Gharaf, Huwaiza, Kumait, Nahr Umr, Noor, Rafadain, Rafadain E, Ratawi and Tuba. Gaddo (1971) reported mean porosities of 9% at Zubair and 16% at Rumaila. Porosities of up to 36% and permeabilities of up to 1560mD occur in rudist rich carbonates. Reservoir thicknesses of several hundred metres and a high net:gross (often>90%) are also recorded. Flurial and Sadooni, 2005 noted that the reservoir quality tends to decrease in downflank wells; this may be due to structural control on reservoir facies or diagententic factors. The Mishrif formation play occupies a relatively narrow belt running NW-SE across central and SE Iraq. Reservoir quality decreases toward the Najaf intra shelf basin to the SW (Gadd0, 1971). The reservoir is massive in the centre of the Mishrif maximum isopach in SE Iraq along the border with Iran (Aqrawi et al. 1998). Toward the flanks internal vertical permeability barriers exist and the reservoir can be divided into separate pay zones. The Mishrif formation is less prospective in the strike-slipe fault zone between Samarra, East Baghdad and Badra. Typical thickness of the Mishrif Formation is 100-170m and is thickest in the Majnoon Field, the Rumalia Field and Zubair Field and near the Iran border, where thicknesses of up to 435m are observed. Reservoir quality is good in rudist reefs, but the reservoir risk is high in Western and Northern Iraq where the Mishrif Formation is thin or absent.

Upper Fars Formation

The formation is typically 600 to 800m thick, sometime called Injana formation along the main Baghdad-Kirkuk road in the NE limb of Jebel Hamrin south. A supplementary type section occurs in well Gilabat-1. The formation is 620m thick.

Euphrates Formation

The shelf carbonate which is equivalent to the Serikagni formation is the Euphrates limestone formation. This formation reaches a maximum thickness of over 50 m at the Jambur structure where it is thickest at the expense of the underlaying Serikagni formation (Al-Juboury et al., 2007). The Miocene Euphrates formation is the youngest unit in the first pay at Kirkuk (Dunningto, 1967). It produces oil at Bai Hassan, Jambur and Naft Khanah (Daniel 1954, IPC, 1956, Dunnington, 1967a; Jassim and Al-Gailani, 2006); it contains heavy at Qasaba, Jawan, Najmah and Qaiyarah (Dunningto, 1967a). Reservoir quality improves toward the top of the formation. The formation may form a reservoir in more basin centre locations than the Kirkuk group, and maybe important in Central Iraq, example at East Baghdad, Balad, Samarra and Tikrit.

The Euphrates Formation consists largely of oolitic to chalky limestones which have commonly been recrystallised. Reservoir limestones were deposited as ramp and slope carbonates, flanked by inner shelf and supratidal carbonates, and limestones are thin and patchy surrounding the ramp and slope carbonates. An area of coarse grained, shallow marine clastics is developed to the southwest, whilst an area of deep basinal carbonates is developed in Blocks 1 and 2. Excellent reservoir quality in limestones is the result of leaching, dolomitisation or fracturing with porosities commonly 10-30% and premeabilities up to several hundred millidarcies. Reservoir potential is good in an area which includes Block 8, together with Blocks 1 and 2, but is poor in the south where the reservoir is thinnest.

Jeribe Formation

The Miocene Jeribe limestone formation with maximum thickness of 63m was recorded at well Injana-5 by Al-Jubouri et al., 2007. Jeribe formation produces oil at Jambur and Naft Khanah (Dunnington, 1967a) and contains heavy oil at Qasab, Jawan, Najmah and Qaiyarah (Dunnington, 1967a), and minor oil in Euphrates graben (de Ruiter et al., 1995). At East Baghdad, extensive sheet-like intervals of anhydrite are interbedded within the reservoir carbonate (Al-Murani, 1986; Sun and Esteban, 1994). In common with the Jeribe formation, the Euphrates formation has reservoir potential in basin centre locations such Tikrit, Balad and Samarra. The main limestone produces oil at the Judaida, Khabbaz, Ajeel and Hamrin fields in north central Iraq, and at Jebel Fauqi, Halfaya and Buzurgan in Southeast Iraq (Jassim and Al-Gailani, 2006). Limestones of the Jeribe Formation are recrystallised and dolomitised and generally massive and were deposited on a broad carbonate platform. Supratidal carbonates and evaporates constitute a non-reservoir facies in lagoonal to reefal conditions. The thickest development of shallow water carbonates is in the centre of the basin and there is a North-South area of low thickness in the Salman Zone. The main basin trends Northwest-Southeast. Shallow marine sandstones and coarse grained continental clastics are developed to the Southwest of the carbonate facies. The reservoir potential of the Jeribe Formation is dependent on secondary porosity development in limestones, together with fracturing and dolomitisation. Low energy, shallow marine limestones therefore exhibit the best reservoir potential and the coarse clastics have poor reservoir potential because they are thin.

Mauddud/Qamchuqa

The formation is mostly dolomites and nertic, detritial-organic limstones and some undolomitiesd, detrial limestones. The Qamchuqa formation to the North and Northeast are producing reservoirs; they are vuggy, fractured and sealed. In central Iraq, fair to good quality sandstones and excellent quality organic limestones occur in palaeo-basin centred around Baghdad. Probable primary porosity is between 7% and 14% with secondary fracturing improving values. In southeast Iraq, toward Basrah, carbonate reservoirs are slightly tectonically fractured and generally well sealed. The entire region to the West of Tel Hajar-1 well in Northern Iraq to Safawi-1 well in the South has no preserved Mauddud Formation.

Hartha

This Upper Campanian-Maastrichtian formation is about 200m thick in southern Iraq, 350m in the northern areas, but commonly much less on the summits of large structures than long flanks. At the type locality Zubair-3 well (from 1704m to 1833m), the formation comprises organic, detrital, glauconitic limestones, locally dolomitised, with grey marls and green shales. Chalky beds are common at Buzurgan, Ghalaisan, Dujaila, Kifl, Musaiyab, Siba and Ubaid; oolitic layers Luhais and Rachi and marly or clayey limestones more common in Buzurgan area. The Hartha potential at Jawan, Majnoon, Falluja, East Baghdad, Najmah, Bai Hassan and Tikrit. had produced 25% oil and 75% water. Oil has been tested at Nahr Umr, Alan, Atshan, Sammara, Kifl, Afaq, Samawa, Ghalaisan, Musaiyib, Makhul, sassan, Ratawi and Rachi. Limestones of the Hartha Formation are commonly interbeded with anhydrite or shale and are strongly dolomitised. Reservoir limestones were deposited in an outer ramp / slope setting, or as shallow water carbonates and have both primary and secondary mouldic and vuggy porosity. There is a small area of mixed shallow marine clastics and lagoonal carbonates in the southwest. The thickness of the Hartha Formation is greatest in the southern part of Central Iraq, and thicknesses >300m are recorded in wells Shawyia-1 and Safawi-1. Reservoir potential is best in the southern part of central Iraq and Southeastern Iraq, where the thickest sequence of shelf and slope carbonates are developed. However, reservoir potential is poor where deep basinal carbonates are developed.

Shiranish Formation

The Upper Campanian-Maastrichtian Shiranish Formation is places directly above the Hartha Formation and may even provide excellent regional seals to its reservoir. Elswhere, the basal part of the Shiranish is age equivalent to Hartha Formation and the top part is age equivalent to Tayarat Formation. Light oil produced from fractures in well indurated, marly planktonic formaminiferal limestones of Shiranish Formation at Ain Zalah and Butmah Fields.

Najmah Formation

The Najmah Formation is largely composed of oolitic limestone which is locally dolomitised. Deposition was in a shallow marine or outer ramp/slope setting, with deep basinal carbonates also developed in the north. The thickness of the Najmah Formation decreases Northwards and Eastwards. This is also reflected by reservoir quality, which is better in the South, the average porosity of the Najmah Formation is 10%.

Nahr Umar

The Nahr Umr Formation is a mixed carbonate/clastic unit consisting of coarse, shallow marine clastics, together with shallow marine carbonates (outer ramp and slope) occupying more distal settings in Eastern Iraq. Sediment was sourced from the Southwest. Deep basinal carbonates were deposited in the most distal parts in the Northwest. Thickness of the Nahr Umr Formation is greatest in the Southeast (up to 360m), but drops to zero in parts of Northeast Iraq. Reservoir potential is best in the shallow marine clastics and the shallow water carbonates, particularly where they are dolomitised, but the reservoir risk is high in northern Iraq where it is thin or absent.

Other Formation

The other formation atr; Serikagni, Sargelu, Shiranish, Hartha, Mushorah, Bitumah, Geli Khan, Dammam, Kifl, Rumaila, Nahr Umar, Ratawi, Sulaiy, Gotina, Najmah and other possible potential formation.

Seals

Information on potential seal rocks was derived from the Oil Ministry Reports on northern and western Iraq, together with Aqrawi *et al.* (2010) and Jassim and Goff (2006). In almost all cases, it is a combination of information from all sources that provided the seal rock distribution and potential. Nine principal seal rock intervals were identified within Iraq. Many maps are showing the seal rock distribution, and seal rock potential based upon lithology, thickness, penetration in wells and proven capabilities in fields. The Base Silurian shale of the Akkas Formation is present over much of Iraq and is a recognised regional seal, and is therefore considered low risk wherever it is present. The Kurra Chine Formation consists of interbedded limestones, evaporates and shales, with the interbedded anhydrite forming the primary seal rock. Secondary potential is also found in tight, anhydritic limestones as well as the Baluty Formation. The formation generally thins to the west and passes into the Mulussa and Zor Hauran Formations in western Iraq. The greatest seal potential is in the northwest, as shown by the proven potential in a number of large fields such as Butmah and Alan. To the northeast and east the seal potential decreases, and in the southern part of the country the limits are uncertain. The Gotnia Formation is a bedded anhydrite with intercalated shales and mudstones and is recognised as a regional seal. The Ratawi Formation is shale dominated with interbedded limestones and marls. As yet, this formation is not proven as a seal in Iraq. The Zubair Formation is comprised of three major sand units separated by sealing intraformational shale units. A tight shale interval at the base of the overlying Shu'aiba Formation provides additional sealing potential for the uppermost sand unit. The Nahr Umr Formation is a mixed clastic/shale unit with carbonate interbeds. The sealing capacity is provided by both intraformational shales (towards the west) and a more continuous Nahr Umr Formation Shale towards the east. The Khasib and Tanuma Formations provide potential seals to the Mishrif Formation. The Khasib Formation is a (marly) limestone which is thickest in southeast Iraq whilst the younger Tanuma Formation is an argillaceous limestone/marl which has a pronounced GR log response which distinguishes it from the Khasib in wells. The Shiranish is a basinal deposit of interbedded shales, marls and argillaceous to marly limestones. It is absent in much of western and south western Iraq (except a thick section in the Anah Graben) where it is replaced by the Tayarat Formation. The Lower Fars Formation is dominated by anhydrites and halites, with thinly interbedded marls and limestones. It is a recognised and excellent regional seal and provides a top seal for many of the Cenozoic (and possibly Mesozoic) discoveries in the Mesopotamian Basin. Elsewhere, the sealing potential is considered good, or very good along a NW-SE belt along the east of the country on account of its considerable thickness (Dujaila-1, 441m) and its proven sealing capacity in fields and discoveries such as Naft Khaneh and Nau Doman.

Production Forecast

The recoverable reserves P50 values have been generated using the recovery factor and oil in place volumes for each lead considered in the blocks. It has been assumed that all leads contain oil or gas. Recoverable oil reserves per well are (12-16) MMstb, recoverable gas reserves per well is (60-120) Bscf. Oil plateau rates and gas plateau rate are variants from field to field. The ultimate recoverable oil reserves per annum are between 7.5-12% and for gas between 7-15%. The oil plateau lengths are (3-8) years and gas plateau length are 3-15 years for different areas. It is assumed that oil wells can be drilled and completed in 30 days and gas wells 40 days. These assumptions are based on the production history of the analogue fields. Recoverable resources based on preliminary deterministic volumetric calculations were used as the basis for production forecasting Table 3.

Zone	Block	Number of Leads	Reservoir	Fluid Type	Reservoir Depth (m)	In Place MMbbl or Bscf	Recovery Factor	Recoverable Volume MMbbl or Bscf
	1	5	Kurra Chine	oil	1700-2300			
			Maudud	oil	1350-1450			
			Shiranish	oil	900-110			
			Others	oil	deeper	7518	0.4	3007.3
			Khabour	gas	3000-4700			
			Akkas	gas				
			Others	gas	deeper	173449	0.8	138759.2
	2	3	Kurra Chine	oil	1050-1412			
			Maudud	oil				
			Shiranish	oil				
			Others	oil		8612	0.4	3444.8
			Khabour	gas	2700-3200			
			Akkas	gas				
Rutba			Others	gas		28443	0.8	22754.4
Zone	3	4	Khabour	gas	2200-2320			
			Akkas	gas				
			Gaara	gas	390-510			
			Others	gas		14690	0.8	11752
	4	4	Khabour	gas	2600-2800			
	&		Akkas	gas				
	5		Basement	gas	6400-6700			
			Others	gas		21321	0.8	17056.8
	6	4	Nah Umar	oil	840-970			
			Mishrif	oil	1600-1700			
			Najmah	oil	1100-1370			
			Others	oil		5342	0.35	1869.7
			Khabour	gas	2600-2850			
			Others	gas		6631	0.8	5304.8
	7	5	Hartha	oil	1300-1360			
			Mishrif	oil	1620-1730			
			Nahr Umr	oil	1790-1860			
			Zubair	oil	1980-2790			
			Yamama	oil	2490-3298			
			Others	oil		2429	0.35	850.15
	8	6	Upper Fars	oil	800-1300			
			Upper Fars	oil	840-1310			
			Jeribe/Euph	oil	2200-3210			
			Others	oil		2234	0.4	893.6
				gas				
Meso				gas				
pota				gas		83511	0.8	66808.8
mian	9	5	Hartha	oil	2250-2380			
Zone			Khasib	oil	2420-2495			
			Mishrif	oil	2520-2600			
			Nahr Umr	oil	3090-3180			
			Shuaiba	oil	3260-3330			
			Zubair	oil	3433-3498			
			Yamama	oil	3850-3980			
			Others	oil		2018	0.35	706.3
	10	4	Mishrif	oil	1800-2200			
			Nahr Umr	oil	1990-2310			
			Zubair	oil	220-2545			
			Yamama	oil	2480-2820			
			Others	oil		2706	0.35	947.1
	11	4	Najmah	oil	1510-1810			
			Mishrif	oil	1800-2200			
			Nahr Umr	oil	1990-2310			
			Yamama	oil	2480-2820			
Salma			Others	oil		3374	0.35	1180.9

Table 3: Recoverable resources based on preliminary deterministic volumetric

The author's estimate based on the known Formation properties of nearby fields which have similar producing reservoirs. The selected main targets are: Formation Kurra Chine Formation which mainly consists predominantly of dolomites and forms the Upper Triassic reservoir in Northwestern Iraq, where it extends into Northeastern and central Syria as a main pay zone. Light oil has been produced from fractured Kurra Chine dolomites in the Butmah and Alan Fields at depths ranging from 2757-3195 m. The gross thickness of the Kurra Chine reservoir in the Butmah Field varies between 85m and 274 m. The reservoir has a relatively low to moderate porosity ranging from 5 to 15%. Predominant pore types are fracture and intercrystalline. Reservoir quality of the Kurra Chine dolomite in the Sufuya Field in Iraq is better than that of the Suwaidiyah Field in Syria, with an average of 20% porosity and an average permeability of 40mD in the Sufuya Field, but an average porosity of only 8% in the Suwaidiyah Field.

The Najmah Formation surrounding Blocks 6, 11 and 12 consists of peloidal-oolitic packstones and grainstones and forms the Upper Jurassic reservoirs for some fields in southern Iraq (e.g., Ratawi, Samawa, and Rumaila Fields). Stratigraphically, it is the lateral equivalent of the Arab-D reservoir which forms the main pay zones in Saudi Arabia, Qatar and western Abu Dhabi. Locally, reservoir quality of the Najmah carbonates was improved considerably by dolomitisation resulting from the reflux of hypersaline brines. Fields close to Blocks 7 and 10. Production forecasts have been calculated based on the assumptions above and are given in Figures 4 and 5.

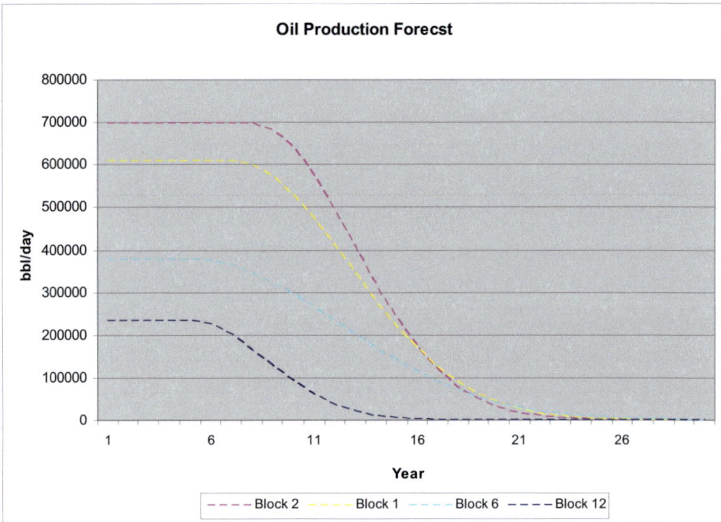

Figure 4: First evaluations summary of production forecast per block

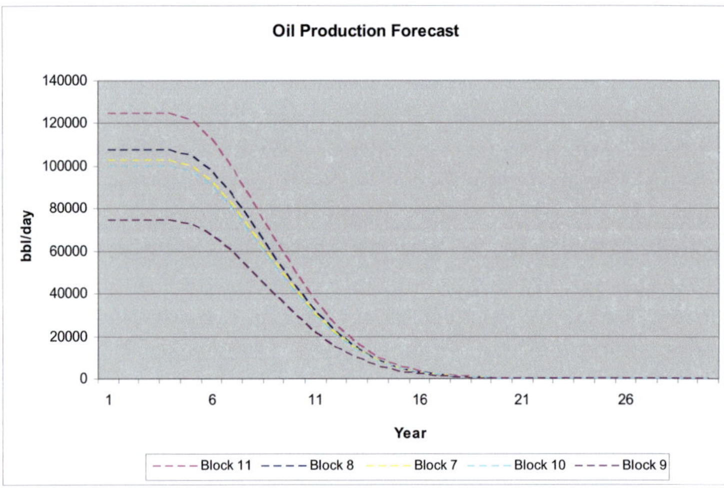

Figure 5: First evaluations summary of production forecast per block

Capital and Operational Expenditures

Several leads were identified in each block with differing hydrocarbon presumption classifications. Generally, all the identified leads were estimated on a 'Stand Alone' basis whereby most of the resources would have their own committed Gas Oil Separation Plant (GOSP). The very large gas resources identified in blocks 1 and 2, it is assumed to be partially used for the Iraqi electricity power generation, petrochemical and private sector basic industry as well the export via Turkey or Syria to Europe or Liquefaction at, for example, Ceyhan or Banias. The cost for the export pipelines to the mentioned ports is matter of agreement between the IOC's and the MoO. In order to reach denunciatory a boe capex, opex, Author created a matrix of costs for the following:

- Geology and Geophysics (G&G)
- General Administration (G&A)
- Drilling (Exploration/Appraisal/development)
- Gas Oil Separation Plant (GOSP)
- Gathering Systems
- Export and Tie-In
- Infrastructure and Camping
- Security Costs
- Indirect Costs
- Contingencies

The costs of the proposed export pipelines, Facilities, gathering systems were estimated using the Que$tor for many benchmark exercises, but mainly using the CERA costs estimation as a basis for the cost estimation.

Drilling costs were estimated from the actual similar wells drilled in Iraq by international service companies. The average cost is $9MM for oil wells and 12 $MM for gas wells. Author didn't consider the cost of drilling oil wells by the own state Iraqi drilling company which is ca. 5.5 $MM and less than this by Chinese drilling contractors. It is known that targets are identified at differing depths and well types may be more costly. Indirect costs were applied at a standard rate of 20% of the base facilities costs. Contingency allowance was similarly applied at a rate of 20% of the base and indirect costs Table 4. Opex boe was determined by estimating the potential life of field of each reservoir. The Opex was declined over the field life using factors and percentages to give an overall life of field (LOF) Opex cost.

Preliminary Cost Expenditure

Block	Prospects/ Lead	Recoverable Resource (MMbbls)	Recoverable Resource (Bcf)	Exploration/ Apprisal MMU$D	GOSP MMU$D	Gathering system MMU$D	Infrastructure MMU$D	Export Pipe MMU$D	Security MMU$D	Sub-Total MMU$D	Indirects 20%	Contingency 25%	Facilities Total	Drilling MMU$D	Grand Total Capex MMU$D
1	5	3007.3		70	680	633	45	88	32	1548	309,6	464,4	2322	1800	4122
			138759.2	240	2100	3100	110	4400	103	10053	2010,6	3015,9	15079,5	10422	25501,5
				10	120	50	8	35	5	228	45,6	68,4	342	0	342
				55	680	672	49	118	30	1604	320,8	481,2	2406	0	2406
				5	28	14	2	25	2	76	15,2	22,8	114	0	114
2	3	3444.8		110	850	862	80	142	96	2140	428	642	3210	2061	5271
			22754.4	200	1600	1900	85	2100	145	6030	1206	1809	9045	1701	10746
				40	410	395	35	114	42	1036	207,2	310,8	1554	0	1554
3	4		11752	80	690	710	65	89	71	1705	341	511,5	2557,5	873	3430,5
				20	85	33	6	43	7	194	38,8	58,2	291	0	291
				20	85	33	6	43	7	194	38,8	58,2	291	0	291
				15	35	17	3	39	3	112	22,4	33,6	168	0	168
4	4		17056.8	90	790	596	60	131	70	1737	347,4	521,1	2605,5	1278	3883,5
&				40	450	393	35	132	57	1107	221,4	332,1	1660,5	0	1660,5
5										0	0	0	0	0	0
6	4	1869.7		50	500	320	40	160	46	1116	223,2	334,8	1674	1116	2790
			5304.8							0	0	0	0	396	396
7	5	850.15		20	200	90	20	32	18	380	76	114	570	504	1074
				10	65	21	5	20	5	126	25,2	37,8	189	0	189
8	6	893.6		15	80	41	6	20	7	169	33,8	50,7	253,5	531	784,5
			66808.8	10	30	9	3	20	3	75	15	22,5	112,5	5004	5116,5
9	5	706.3		15	65	33	6	12	6	137	27,4	41,1	205,5	423	628,5
				15	75	26	7	12	6	141	28,2	42,3	211,5	0	211,5
				10	20	6	2	6	2	46	9,2	13,8	69	0	69
10	4	947.1		20	130	65	10	30	10	265	53	79,5	397,5	567	964,5
				15	80	25	7	25	5	157	31,4	47,1	235,5	0	235,5
				15	85	29	7	28	7	171	34,2	51,3	256,5	0	256,5
11	4		9336	120	1150	490	85	105	97	2047	409,4	614,1	3070,5	702	3772,5
		1180.9		50	500	320	40	160	46	1116	223,2	334,8	1674	702	2376
12	4	2212.7		50	500	320	40	160	46	1116	223,2	334,8	1674	1323	2997

Table 4: Preliminary cost estimation

Fiscal Terms

The main terms of the 4th bid round are; all production is owned by the folk of the Republic of Iraq and the contractor wear the risk and cost of exploration, development and production. In return for exploring, developing and producing from the block, the contractor is re-imbursed his costs and also receives remuneration based on the bid fee in US$/boe submitted by the contractor. The state partner takes a 25% carried interest, the state partner wear no cost but receives 25% of the remuneration. The maximum contract term is for 30 years excluding exploration extensions, whilst the maximum development period is 25 years. Petroleum costs accrued from the effective date are payable from 50% of the unexpended revenue. The remuneration is payable from 20% of the unexpended revenue less the recovered costs. The applicable remuneration fee bid is scaled down according to the calculated R factor (cumulative revenue over cumulative costs), as per the table 5 illustrated. There is no remuneration when the net production below commercial prod rate. Additional terms include, Infrastructure fund payable of 10% of the contractor's annual budget, training fund payable of a minimum of US$1 million per year, signature bonus payable within 30 days of effective date, contractor income tax at a rate of 35%.

R Factor Range for Remuneration Fee

From	To	% of RFB
0.00	1.00	100
1.00	1.25	60
1.25	1.5	40
1.5	2.00	30
2.00	100.00	20

Table 5: Remuneration fee adjusted by R factor

Author assumed on the terms of the draft contract are; remuneration fee of US$5/boe, commercial production rate = zero (impacts R factor remuneration), signature bonus payable US$50 million. Furthermore the fund for decommissioning was set up 10 years backward to the end of life of field, containing equal payments over the 10 years. These are considered recoverable costs. The economic assumptions are:

- The assessment date is the 1st January 2013
- Assumed discount rate is 10%
- Brent Oil Price US$89/bbl. Assumed 2013 constant real, inflated 2% per annum from 2013
- Gas Price of US$ 3.48/mcf as per contract 25% of export oil price, assumed 2013 constant real
- NGL Price US$ 50.16/bbl as per contract 60% of export oil price, assumed 2013 constant real
- Contract: 1 boe = 1 bbl of crude oil or LPG = 6,000 scf of dry gas
- Mid-year discounting of cashflows
- Annual cost inflation of 2%

Contract - Participants

- Ministry of Oil (MoO)
- Own State Regional Oil Companiesy (ROC)
    - State company granted exclusive rights to Contract Area
- Oil Exploration Company (OEC)
    - Assigned rights and obligations during de-mining and exploration periods
- Contractor
    - IOC / consortium + Iraqi state partner

Contract - Other Terms

- MWO to be completed within first 4 years of Exploration period
- De-mining starts from approval of exploration plan
  - Can last up to 1 year
- Exploration start after de-mining period
  - 4 years with possible 2 year extension
  - Further 2 year extension possible to complete appraisal of a discovery
- Development period starting from the approval of the plan of development
  - 20 years with possible 5 year extension
  - Maximum contract term of 30 years (excluding extensions)
- ROC may approve a declaration of commerciality but require a holding period of a max of 7 years prior to development
- Contractor may choose to receive reimbursement for allowed pre-development costs + interest at LIBOR + 5%, and petroleum costs

Economics

The principal economics calculation under the above mentioned Fiscal Terms is as following:

Contractor Cash flow

Deductions

Project Costs
- De-mining costs
- Exploration Costs
- Development Costs
- Operating Costs
- Decommissioning

PSC Costs
- Signature Bonus
- Infrastructure Fund
- Training Fund
- Income Tax

Contractor Income

\+ Cost Recovery
\+ Remuneration

Contract - Remuneration

- Accrued and payable in the quarter of first commercial production
- Remaining Production * Applicable Remuneration Fee
  - Remaining Production (BOE) = Net Production – Cost Production
  - Applicable Remuneration Fee = Remuneration Fee Bid * R Factor
  - No remuneration when adjusted net production below commercial prod rate
- Payable from 20% of remaining revenue
  - Remaining revenue = remaining production * provisional export oil price
- Remuneration Fee from R Factor table US$/boe

R Factor Range for Remuneration Fee

From	To	% of RFB
0.00	1.00	100
1.00	1.25	60
1.25	1.5	40
1.5	2.00	30
2.00	100.00	20

The remuneration is payable from 20% of the deemed revenue less the recovered costs. RFB is scaled down according to the calculated R Factor (cumulative revenue over cumulative costs), Income Tax of 35%. Petroleum costs are payable from 50% of the deemed revenue. Contractor is re-imbursed costs and receives remuneration based on the Remuneration Fee Bid (RFB) in US$/boe. The remuneration fee is variants from lead to other. The Net Present Value (NPV) for the P90, P10 and P50 cases were calculated. The Annual discounted cashflow model was created, taking in consideration the economic assumptions, the contract terms and NPV's cases. The calculated Internal Rate of Return (IRR) is 15% and the range of the NPV's is in some blocks in minus value and some others reach to more than 1900 MMU$D per lead.

Conclusion

The reservoir volumetric and quality of geological units discussed here represents a synthesis of more detailed work carried out. The stratigraphy and reservoir quality from neighbour fields and the extent of the quality of these elements were analysed. The preliminary evaluation is a generalised one and will need to be refined against global analogues and for the specific depths etc. of individual prospects. For details, it should be more data required for individual sections. Details of reservoir are characteristics of various fields but most usually fields around the exploration blocks.

It is clear from the available data, that detailed seismic mapping can produce a much more accurate play model for specific areas. Most of the results of the seismic are poor which include Western and South western to the South of the country; it is highly recommend to shoot 3D seismic. This article is based on regional data, therefore it is likely that many details that are important locally, e.g. absence of formations off individual fault-block highs not penetrated by wells, or presence of units in basins not tested by wells, will have been overlooked. Where possible the likely variance involved in these smaller-scale interpretations is highlighted, but precise interpretations cannot be guaranteed.

Old wells and generalised stratigraphy of northern Iraq have been summarised in a number of publications. Stratigraphies of the older wells are relatively well known, particularly since the Iraq Petroleum Company (IPC). Post-IPC wells, such as Jebel Khand-1 and Abtakh-1, were mostly drilled with assistance of Eastern European countries and their stratigraphies need to be treated with caution. Also, tying seismic to well data is presently difficult because of the opaqueness of poor-quality seismic lines over the tightly folded anticlines on which most wells are situated. Lines shot between the surface anticlines often reveal the best data quality, although even in these the presence of numerous anhydrites and halite in the stratigraphy, plus karstic features at surface, also serves to dampen seismic energy and its penetration. Another issue that strongly affects interpretations of parts of the stratigraphy is the confidence that can be placed in published palaeontological and lithostratigraphical interpretations.

The prospectivity of northern Iraq is significant. It sits on margin of two known hydrocarbon provinces, firstly the Mesopotamian basin with its major Jurassic source rocks and Cretaceous to Oligo-Miocene reservoirs in late Tertiary anticlinal traps to the SE (Kirkuk-Makhul-Hamrin area) and secondly, the ‘NW Iraq/NE Syria’ hydrocarbon province with likely Triassic source rocks and Triassic-Middle/Late Cretaceous reservoirs in inverted half-grabens to the north. The Khleisia High is largely unexplored, but possesses many strong geological similarities with the Euphrates Graben system of Syria, in which Silurian, Triassic and Late Cretaceous source rocks charge tilted fault-block traps in which reservoirs, consisting of a variety of lithologies, range in age from Carboniferous to Neogene. It is reasonable to assume that further accumulations remain to be discovered within the 4th bid round twelve blocks. Until seismic mapping, volumetrics and play/prospect risk is assessed, it is difficult to put a precise figure on yet-to-find.

In all blocks there are indications of oil, condensate or gas. Some of the exploration wells are tested in some intervals and some of the wells have oil and gas shows. The same formation in the exploration blocks produces oil or possible gas in the neighbour fields. The selected main reservoirs appear to have locally good reservoir quality. In some of the reservoirs are depositional heterogeneities and interbedding with the other formation results in complex reservoir geometry with variable thicknesses. In some intervals are somewhat more or less porous and permeable. These are isolated from each other, with different oil column heights and independent fluid levels and different oil gravity.

The economic calculations show that the development of separate leads could bring a positive and in some leads negative results. It is recommendable to develop many leads together in the same block, so that could increase the NPV of the intended investment. Increase the RFB % could make the NPV's more attractive for the IOC's.

**the paper submitted by the Oil Gas Journal (OGJ, 10/04/2011. Detailed analysis is available by the Author.*

References

1. Oil Ministry Report, 2001

2. Iraq Oil Ministry 4th Bid Round, 2010

3. Aqrawi A.A.M., Goff J.C., Horbury A.D. & Sadooni F.N. 2010 The petroleum geology of Iraq. Scientific Press 424pp.

4. Country Overview Iraq Middle East - NW Arabia Upstream Service December 2010 Page 4 of 5

5. Jassim and Goff, book

6. Al-Ameri, T.K., 2000, Silurian palynostratigraphy and the unconformable boundary with the Devonian, western Iraqi desert. Iraqi Journal of Science, v.41c, p. 119-137.

7. Al-Ameri, T.K., and Baban, D.H.A., 2002, Hydrocarbon generation potential of the Lower Palaeozoic succession, western Iraqi desert. Iraqi Journal of Science, v.43 no.3, p. 77-119.

8. Abawi, T.S., 1989, Foraminifera, stratigraphy and sedimentary environment of the Euphrates Formation, Lower Miocene, Sinjar area, northwestern Iraq. Newsl. Stratigr, v21, pt. 1, p.15-24

9. Oil Exploration Company Report, 2001

10. Muhammed Mazeel, Siba Gas Field Development Plan: Subsurface Uncertainties And Economic Criteria, Vol 53, No. 27, MEES

11. Aqrawi et al. (2010) and Jassim and Goff (2006)

12. Muhammed Mazeel, Iraq's Technical Service Contracts – A Good Deal For Iraq?, Vol 53, No. 3, MEES

13. Muhammed Mazeel, Iraq's Ahdab oil field development limits contractor profitability, Oil Gas Journal, 01 August 2011

14. Muhammed Mazeel, Hydrocarbon potential, reservoir development estimated for western Iraq's Akkas gas field, 01 October 2011

The author

M. Abed Mazeel (m.mazeel@hotmail.de) has been employed in many European oil companies as a senior reservoir/ petroleum engineer. He has also worked as director general of Iraqi Drilling Company and director general for Oil Products Distribution Co. He was an oil industry adviser to Iraqi Prime Minister Ibrahim Al-Jaafari. M. A. Mazeel studied at the Faculty of Mining and Geology-Department of Petroleum Engineering of the University of Belgrade (Master in engineering), the Escuela Superior de Minas in Madrid (credit in applied geophysics engineering) and the University of Clausthal-Zellerfeld in Germany. He holds a PhD in petroleum engineering by the Technical University of Clausthal and is currently working on a PhD in petroleum economics. He is currently working as Subsurface Director at International Oil Company. His next books cover topics such as Gas Production Engineering and Iraq Fiscal Regimes.